A Century of Aerospace History

C. Wayne Ottinger

Archway Publishing books may be ordered through booksellers or by contacting:

Archway Publishing
1663 Liberty Drive
Bloomington, IN 47403
www.archwaypublishing.com
844-669-3957

Because of the dynamic nature of the Internet, any web addresses or links contained in this book may have changed since publication and may no longer be valid. The views expressed in this work are solely those of the author and do not necessarily reflect the views of the publisher, and the publisher hereby disclaims any responsibility for them.

Any people depicted in stock imagery provided by Getty Images are models, and such images are being used for illustrative purposes only. Certain stock imagery © Getty Images.

Cover Credits:
Author's Concept Implemented by Angela Saylors, Graphic Artist

ISBN: 978-1-6657-4903-9 (soft cover)
ISBN: 978-1-6657-4904-6 (case bound)
ISBN: 978-1-6657-5037-0 (dust jacket)
ISBN: 978-1-6657-4905-3 (e)

Library of Congress Control Number: 2023915892

Print information available on the last page.

Archway Publishing rev. date: 09/25/2023

Credits

"In Wayne Ottinger's book, readers are taken on a journey through the life of a remarkable individual who was a key player in the beginning of the US adventure to Space. With skillful anecdotes and meticulous research, Ottinger delves in the highs and lows, the triumphs and challenges, that shaped the aerospace journey at its beginning to include the X-15 to the Lunar Lander Research and Training vehicles. This book not only offers a view in the life of its author, but also explores universal themes of perseverance, passion, and the indomitable human spirit. A must-read for anyone seeking to understand how one key individual participated in the early chapters of aerospace."

John L. Barry, Major General, USAF (Ret)

"Wayne Ottinger has produced an always fascinating personal review of aerospace history as lived and witnessed by a practicing and highly creative engineer. He strips much of the myth and mystique from both the aerospace design process and the flight test enterprise, and, in so doing, casts new and provocative interpretations and insights that add significantly to our knowledge. Highly recommended!"

Dr Richard P Hallion FAIAA FRAeS FRHistS

"Wayne played a key role in some of the most critical research, development, and flight test programs in the history of The US Space Program, from the X-15 to the Lunar Lander Research and Training Vehicles. His creativity, innovation, and hard work on these development programs directly contributed to the successes of both the Apollo moon landings and the eventual development of the Space Shuttle. If you're interested in learning about how we figured out how to land on the moon, then this book is for you."

Steve Lindsey, Col, USAF (ret)
NASA Astronaut & Space Shuttle Commander (62.9 days In Space)

"Wayne is a walking encyclopedia of aerospace history and knowledge. His book is not only a compelling summary of his illustrious career, but a valuable reference for those interested in the development of space related vehicles and systems."

Joe Tanner, retired shuttle astronaut

"I enjoyed finding some stories overlapping my great flying days at NASA Flight Research Center in Wayne's detailed capture of many programs."

Fred Haise NASA Research Pilot Apollo 13 LMP Enterprise CDR

"Wayne Ottinger's look back over a long life in critical periods of aerospace history represents a monumental contribution to the nation's record of success in this arena."

Harrison H. Schmitt, Apollo 17 Astronaut, United States Senator, Earth and Aerospace Consultant

Contents

About The Author

My motivation to write this book is to share lessons learned and a substantial number of rare pictures that help tell the stories of the many aerospace and high-tech projects he carried engineering and management roles for seven decades and connections with his collogues for three decades earlier. I hope current and future generations of aerospace and high-tech engineers and management will benefit from applying the lessons learned from these stories of the struggles and initiatives taken to recover and adapt them according to their experiences. I hope this book's value will avoid the unnecessary replications of mistakes made throughout history.

The unique book design uses the ten-decade theme to provide the reader with my extensive professional career. In addition, the five Appendices display about 200 pictures, half considered rarely available, making this book a valuable resource for present and future generations. Appendix D is a transcript of notes from Dave Scott's comments from a NASA conference I initiated in December 2009.

I was born on October 22, 1933, in Phoenix, Arizona. I witnessed a combination of depression and WWII in my early years as a younger brother to his six years older sister. My father, a school teacher, and neighbor, Mrs. Hickey, born in 1863, provided an environment forming my personality and respect for elders. My Victory Garden began the motivation driving my work ethic prominent throughout my professional career. I married at 18 at the end of my first year of college, and became a father at age 20. This early start set the stage for my full-time work as a full-time student, graduating with a BSME in 1955 from the University of Arizona. In High School at 16 in 1949, I soloed in an Aeronca Champion, obtaining my private pilot's license on May 26, 1951.

In September 1955, I worked at the General Electric large engine plant in Evendale, Ohio. I spent long hours on extended shifts testing the new J79 Preliminary Flight Rating (PFRT) engine, focusing on the afterburner and performance aspects before its first manned flight in December 1955 in Douglas X4D. In June 1956, I arrived at Edwards AFB working on early J79-powered F-104s, B-58s, Regulus II missiles, and a Grumman F-11-1F.

1960 I was the X-15 flight operations engineer responsible for the propulsion systems for three years. In 1963 he served as the NASA Lunar Landing Research Vehicle (LLRV) project engineer until I joined Bell Aerosytems as the Lunar Landing Training Vehicle (LLTV) Technical Director in June 1966. I was responsible for adding several Apollo hardware systems to the LLRV design and delivering three LLTVs for training The Apollo Commanders and Backup Commanders.

In the 1970s and 1980s, I performed high-tech engineering and operations services for various industries, including Air Cushion Vehicles, Gas Centrifuges for Uranium Enrichment, and Solar Energy. In the 1990s, I founded a nonprofit for educating aerospace history. For the next two- decades of the 21st century, I served as a NASA senior consultant and participant in many national aerospace conferences.

Acknowledgments

To my Family and Boulder Colorado Friends, My two years of starts and stops and expansion of the vision of the stories and realization of how rare many the pictures have received encouragement and patience from all, much appreciated.

Jim Love, NASA DFRC Operations Engineering Branch Chief (later the X-15 Project manager): Jim offered me my first NASA job at 27. I worked with Milt Thompson in 1956 as a GE propulsion flight test engineer. They had just received their J79-powered F-104s. Jim put me through a series of challenges that I look back on as a critical role in maturing rapidly in ensuring flight safety while finding root causes of hardware and system failures.

Don Bellman, NASA DFRC LLRV Project Manager:

Don had initially considered a newly discharged Air Force Pilot to be the LLTRV Project Engineer. Perry Row, who had replaced Jim Love as Flight Operations Branch Chief, offered me the LLRV or Dyna-Soar X-20 Project Engineer positions. I asked for the LLRV. Don was satisfied I was a good choice. Don and I worked together well to achieve the budget constraints imposed by Paul Bikle. After delivery, we solved them by shifting the final assembly, electrical, and rocket system to my crew at Edwards.

Paul Bikle, NASA DFRC Director:

My first flight safety engagement with Paul was the X-15 persistent rocket engine in-flight small explosions of a steel pressure transducer tubing on several flights. Initially, I asked Paul to ground the X-15s until we could determine the cause. He denied my request but allowed me to have the machine shop fabricate temporary steel shields around the tubing where the explosions occurred. The Reaction Motors field reps blamed the explosions on an incorrect diagnostic. It took four or five months for the Air Force Power Plant Branch to tear down all eight flight engines to replace the LOX Safe lubricant used in the engine's plumbing. After the long delays, one of the engine's firing tests in the test stand blew it up. A vibration problem was the final root cause. Paul always assured me I was in charge of the Flight Operations we had to conduct at the South Base, an eight-mile trip by road. This arrangement was a deviation from normal center operations.

Dr. Richard (Dick Hallion) USAF Chief Historian Emeritus:

In 1994, I approached Dr. Richard (Dick} Hallion, Senior Historian of the USAF at Bolling AFB, to explore a grant opportunity under a new Department of Defense (DoD) Cold War History program. The project I proposed was to document the NASA Lifting Body program from the early 1960s. He was accommodating, guiding me through the process of securing DoD funding. The funding launched my first nonprofit, which served NASA education and Intergovernmental Personnel Act (IPA) projects for almost eight years. His tenure as the Historian at the Air Force Flight Test Center at Edwards AFB, and as recent as today, continues our relationship.

Dr. Harrison (Jack) Jack Schmitt, Apollo 17 Astronaut:

Jack was Chairman of the NASA Advisory Committee during the 2 or 3 years as we were planning the Go For Lunar Landing Conference. Neil Armstrong and Mark Robinson, Principal Investigator (PI) for the NASA Lunar Reconnaissance Orbiter Camera) were also on the Committee. Jack chaired the 2008 Apollo Go for Lunar Landing Panel and participated in NASA's JSC conference on December 9, 2008, which I had promoted, where he joined Neil Armstrong, Gene Cernan. and John Young, hosted by the Constellation Altair Project Office. Jack was accommodating in assisting in the 2008 events.

John Kelly, NASA AFRC Flight Opportunities:

John attended the Go For Lunar Landing Conference in Tempe, Arizona, in March 2008. As a result of that conference, John was the NASA DFRC (before the AFRC name change) team leader in delivering a trade study to determine a new Altair Lunar Landing Training Vehicle. I supported John as a NASA Shuttle & Apollo Generation (SAGE) consultant for two years and appreciate his continued support.

Dr. John Tylko, Aurora Flight Sciences Chief Innovation Officer:

In June 2017, John was in Denver at an AIAA conference and called me to visit Boulder to delve into my aerospace history. We spent several days getting acquainted and then left for Tucson to attend the Spacefest. I participated in an X-15 panel with Astronaut General Joe Engle. John kept in touch and had to reschedule an AIAA Engineering Apollo Panel from 2019 to January 2020 in Orlando. John has been teaching the Engineering Apollo MIT Graduate course and chaired our panel with Dave Scott, Frank Hughes, and myself. Several months after, he asked me to consult on a team he had put together at Aurora Flight Sciences in Cambridge, MA, to perform a proposal for an In-Flt Trainer for the Artemis Lander. For about 18 months, we worked virtually, developing design concepts with new electric propulsion technologies. Unfortunately, the delay in downsizing the three NASA lander contractors and NASA's autonomous Space X choice extinguished interest in any trainer. Nevertheless, John has been good to me and appreciated.

David Dean. Smith, My Neighbor

Dave's help in editing this manuscript is much appreciated. His work as a Mechanical Engineer for Motorola from 1984 to 1987, producing technical manuals for their Space Shuttle Control system components, were skills fortunate for my need to submit this manuscript to the Archway Division of Simon-Schuster.

Foreword

One quiet morning during the early dawn of the digital age, the first-ever fly-by-wire" flight vehicle lifted off from the calm high desert of the Flight Research Center (FRC) at Edwards Air Force Base in California. They called it the *"Lunar Landing Research Vehicle"* (LLRV). Then after many successful test flights, it was to be known as the *"Lunar Landing Training Vehicle"* (LLTV). The day was in 1964. Five years later, in July 1969, after twenty-one flights in the LLRV and six flights in the LLTV, Neil Armstrong made the first manned landing on the Moon in the Apollo Lunar Module. The LLRV was the first airplane to fly, relying entirely on a fly-by-wire control system eliminating the complexity and weight of hydraulic and mechanical control systems. The first airplane to be controlled by digital Fly-by-Wire did not occur until eight years later, in May 1972. Today, every contemporary aircraft is controlled by fly-by-wire avionics, long past the ancient cable-pully and hydraulic systems of the early 20th century.

This remarkable machine was to be the foundation of the first six human landings on the Moon, each of which was the result of the pilot in command (Commander) visually selecting the touchdown point and being solely responsible for controlling the Lunar Module (LM), from guidance by a small slow computer to a manual landing by hand to a precise target location on an unknown irregular surface of boulders and craters in a 1/6 G vacuum environment with only minutes of propellant remaining before the landing had to be successfully completed or an abort was required.

As Pete Conrad, the Commander of Apollo 12, once said:

"NASA is betting the whole program on a guy who does not mess up his first landing."

Only one chance to land, no opportunity to go around and try another approach, no opportunity to extend the hover time and fly around the surface seeking the best or a better landing point, and certainly no opportunity to go back to orbit and try again. No landing, and the mission was over. The pilot in command alone must not only make the final decision, but with a limited amount of time available, he must also be able to maneuver a unique rocket-propelled flight vehicle in an airless, low-gravity environment with no assistance or communications from Mission Control whatsoever. He alone must be prepared and be confident in his abilities to complete this challenging task – without failure…!!

The final landing profile for the lunar descent begins at 7,500 feet above the surface, flying at 300 mph and 6.5 miles from the target when the LM pitches forward for the first view of the landing site by the crew. With only 4 minutes and 30 seconds of propellant remaining, the pilot (Commander) must select a landing point in an area of craters and boulder field acceptable to the LM structure (strength, angle, and rate of "tilt"). Until the pilot takeover at 400 feet, the LM was controlled automatically by a single small, slow computer. Should the computer fail at any point, the pilot had to take over and fly the vehicle with a unique control system installed in the LM similar to an aircraft pilot control system (throttle and hand controller [stick]).

And how would the pilot achieve the flying proficiency and confidence on the Earth, long before he got to the Moon? The answer: The LLTV (LLRV), which was invented, tested, and verified by an exceptional group of pilots and engineers at the Flight Research Center, Edwards AFB, California. And thank goodness it was – the six Apollo landings on the Moon by humans were each 100 percent successful, marking a major milestone in the history of human flight and the history of the Moon.

Soon after the Apollo Lunar Module was designed and the lunar landing flight profile was defined, the engineers and pilots at the FRC were given a daunting task. It was recognized that no existing flight vehicle would prepare a pilot for a manual landing on the Moon – the performance and handling qualities of the Lunar Module were vastly different from any known manned flight vehicle. And these flight characteristics had to match the same as the Lunar Module when landing on the Moon. However, the LLTV was a single-pilot training vehicle in which a local Mission Control Center monitored most systems.

The LLTV control system was necessarily required to match the LM control system in both operation and response or performance and handling qualities. In the cockpit, this system consisted of a throttle to control the main engine and a hand controller to control the vehicle's attitude. The main engine thrust was fixed vertically, meaning that the lateral and vertical motion of the vehicle were tightly coupled – the throttle controlled the thrust of the engine, and the vehicle attitude controlled the resulting "tilt" of the vehicle - increased thrust would result in both vertical and lateral motion, i.e., the vertical and horizontal motions were very tightly coupled and to achieve the desired motion, due to the natural delay in response to the hand controllers, the pilot had to anticipate the slow response and instantly input the desired control inputs. One analogy to this maneuvering challenge was like attempting to change direction on an ice rink wearing dress shoes – think about it…!!

And before the LLRV had emerged, none of these landing maneuvers or LM flight characteristics had ever been tested in actual flight conditions. So the engineer/pilot team began with a blank piece of paper to not only test and demonstrate LLRV capabilities but ensure they matched the flight capabilities of the Lunar Module – but fortunately, the Flight Research Center personnel were the most experienced in the industry. As a result, they accepted this challenge with vigor, insight, and innovation.

After the LLRV had been tested and approved for astronaut training, the engineer/pilot team had to develop a training program for astronauts who would fly the machine, a training program that upgraded their piloting skills to land a unique flying vehicle successfully under the above conditions and LM capability – successfully meaning before propellant depletion and at a visually selected point on the unknown lunar surface that complied with LM landing capabilities. For practical training conditions on Earth, it was determined that the most vulnerable point in the Lunar Module descent was 400 feet above the surface – a point at which the remaining propellant would enable flight for only about four more minutes.

I began my LLTV career in March 1969 as Backup Commander of Apollo 12. Followed by Commander of Apollo 15 in November 1969, and after 32 flights in the LLTV, my LLTV career ended only one month before the launch of Apollo 15, July 1971.

Looking back at my descent and landing at the Hadley Apennine site on the Moon, I realize that because of the very effective LLTV training and the skills of the engineers and pilots who invented the LLTV, when the LM pitched over. I first saw the lunar surface and was clearly "in the zone." I was able to fly the LM intuitively as if it were actually part of me, I did not need to think about any control inputs, either by throttle or hand controller. The LM just did what I thought it should, with no need for conscious control inputs. And significant maneuvers were required because initially, I could not locate our intended landing point -- I had to correct for a 3,000-foot error in the guidance trajectory to our landing point. But when you are in the zone, time slows; and without the LLTV training, there would not have been enough mental time to maneuver the

LM to an acceptable landing point near our original target before running out of sufficient propellant for the final descent and touch down. But prior to the descent, based on the very effective training in the LLTV, I was very comfortable in anticipating the most difficult landing of my career.

Yet perhaps the most amazing aspect of the LLTV was the manner in which its performance and handling qualities matched those of the actual Lunar Module that I landed on, Apollo 15 (LM-10). Flying the LM manually was almost identical to flying the LLTV manually. Controlling the LM dung the final 400 feet to touchdown was "just like home" -- the LM hand-controller moved the LM in exactly the same manner the LLTV hand controller moved the LLTV. And throughout the final descent from 400 feet, the amount of flight time remaining (propellant) in the LM before touchdown (abort) was almost identical to the amount of flight time in the LLTV during any single training flight. The preflight preparation of the Commander (the pilot) had been suburb – and that is why I was "in the zone" during my lunar landing – thanks to the FRC pilots and engineers, another chapter in your book of exceptional flight vehicles.

The post-flight reports from each of the six Apollo missions regarding the success of a one-time attempt to land on the Moon confirmed in the opinion of each of the Commanders

(pilots) that the LLRV was absolutely mandatory to train a human pilot to land a vehicle (the LM) of the Apollo era successfully on the lunar surface.

Wayne Ottinger and his colleagues at FRC had created a marvelous flying machine called the Lunar Landing Training Vehicle, a unique machine that gave us the confidence and capabilities to ensure the Apollo Lunar Module never failed in landing at six different and unknown locations on the Moon. This remarkable book tells the full story of how they did it – Apollo will be ever thankful for their inspiration and innovation, a true milestone in the history of human flight.

David R. Scott Commander, Apollo 15
March 3, 2023

Take a low fly-by of the last century of my aerospace connections and explore these stories and pictures.

The 1920s: General Doolittle setting air race records. I worked with his grandson in 2002. My boss in 1969, General Schweitzer (1906 - 1996), flew in 1929, helping Chiang Kai Shek from 1932 to 1935 start his Air Force in China. He worked as a pilot for American Airlines in 1936.

Hannah Reitsch, M.G. Bekker, Richard E. Day, Betty Love, Whitey Whiteside, and Robert A. Hoover

The1930s: I was born October 22, 1933 in Phoenix Arizona

The 1940s: At 16 in 1949, I flew solo in the high school flying club. In 1942 Ed Rhodes was the Bell Aircraft XP-59, the US first jet airplane project engineer. He later worked for me in 1966 as we updated the Apollo Lunar Landing Research Vehicle with new Apollo Lunar Module hardware to build three new Lunar Landing Training Vehicles for astronaut training.

The 1950s: I worked full-time outside jobs for one year of community college and three years at the University of Arizona, BSME 1955. I started a family of 4 sons in 1954. My aerospace career began in June 1955 as a GE test engineer and progressed to a jet engine flight test engineer.

The 1960s: The Apollo Decade:

As a NASA flight operation engineer, I'm proud of avoiding significant program delays and prohibitive expenses by designing a pantograph machine to enable a one-week field repair of the XLR 99 60,000 lb. thrust rocket engine's chamber ceramic coating. In addition, just after joining NASA in April 1960, I observed Eisenhower's attempt at the NASA Flight Research Center taxi demonstration that Gary Francis Powers U2 spy plane was on a NASA mission.

August 1962- 1966 NASA Lunar Landing Research Vehicle project engineer participating in the design, building, and ground and flight testing of the first aircraft to use Fly-By-Wire flight control systems in 1964. The pilot training requires flying the last three hundred feet of the lunar landing trajectory in a one-sixth-G simulated environment without any perceived aerodynamic forces affecting the vehicle dynamics. These imposed the test pilots and astronauts to learn on an earth-bound flying VTOL (Vertical Take-Off and Landing) aircraft to perceived aerodynamic forces affecting the vehicle dynamics. The most valuable training was using 5 to 6 times the amount of pitch and roll attitude angles used in helicopters flying on earth. Our 30 months of 200 research flights provided significantly improved pilot training qualities at lower control authorities, enabling Grumman to abandon VTOL handling standards and save the weight and complexity of separate control systems for a heavy landing mode and a lighter accent mode.

June 1966 - September 1968: The Bell Aerosytems Lunar Landing Training Vehicle (LLTV) Technical Director. NASA contracted with Bell to build and test three LLTVs with upgraded flight control components provided for the Grumman LM giving the astronauts identical controls for training and space operations. The LLTVs had partly improved training capabilities by deleting hardware from the LLRV design used for the flight research program. After the January 1967 Apollo fire, a critical two-month delay in deliveries to NASA was resolved by a one-week structural fatigue test on the aluminum welding of a complex landing leg. This test allowed the NASA inspectors to accept the Bell design experts' X-Ray interpretation of minor flaws they were worried about. I used ten times the test loads required. Neil's LLTV training was only one month before the Apollo 11 launch in July 1969. Stopping the two-month delay makes me proud.

1969: After the late 1968 return from Bell's medical leave from the Apollo program, I was assigned to work in the Los Angeles office for Brigadier General John M. Schweizer Jr. (USAF Ret). He headed the West Coast Marketing Office. From 1932 to 1935, General Schweizer was an aviation advisor and flying instructor for the National Government of the Republic of China as a member of an American group that went to China to organize and train an air force for Generalissimo Chiang Kai Shek. I was privileged to hear his stories as I worked and traveled with him, working on marketing Bell's Air Cushion Vehicle (ACV) and electronic product lines with DoD and commercial markets. The Alaska North Slope Oil discovery and use of ACVs for ground transportation in arctic tundra for the petroleum industry inspired me to leave Bell and form the Transport Development Company. Bell had six 25-ton ACVs they had refurbished from their Viet Nam river combat tours. Based on my contacts with prominent petroleum executives, I felt I could provide ACV support for their exploratory North Slope operations. By mid-summer, a Canadian 25-ton ACV had been operating for ARCO. A fatal Air Crane crash followed by a fatal ACV crash killing an unseated passenger as it hit a river bank at a high speed inspired me to design a lower-cost wheeled ACV.

October 27, 1969: the Apollo 11 crew, a week before, had finished their three-week quarantine, and Robert A. Hoover, President of the Society of Experimental Test Pilots, invited them to the Los Angeles Beverly Hilton Hotel for the annual symposium banquet. Appendix A's rare pictures show Charles Lindberg, Mrs. Hoover, and Neil Armstrong at the table. This picture ranks near the top of the book's many rare of about 200 pictures.

The 1970s

My connection with Joe Fletcher, who I met at the Rand Corporation in Santa Monica (stories told with pictures), helped me launch my company and get going in Alaska. Stories for the next few years tell of close calls on acquiring product development funding, trips to the Pentagon meeting with the Director of Defense Research and Engineering, and Vice Admiral Forest S. Petersen (who I had worked with when he was an X-15 pilot). In 1977 I extended my ACV experience by performing engineering consulting services for Rohr Industries on the 100-ton SES-100 and the 3,00 ton SES-3000 100-knot Surface Effect Ships. 1977-1988 California Energy Commission: I produced films an organized renewable energy and energy conservation for municipal utilities.

The 1980s: 1980 - 1988: Owner-built Passive Solar House of a 2,400 sq. ft., all-electric house in Vista, CA. Stories were published in Sunset Magazine and San Diego newspapers for the $35/mo total energy costs.

1979-1987: Garrett AiResearch I developed imaging systems for failure diagnostics. These included advanced fiberscopes, borescopes, and high-speed laser strobe lights for high-speed video vacuum spin tanks for uranium enrichment gas centrifuges.

1988-1990: SynerVision: My new company offering image services to industry and two nuclear powerplants using tools acquired in a previous job working on uranium enrichment gas centrifuges

The 1990s:

1990-1993: Garrett Engines, Phoenix: Flight Control engineer on the Rockwell OV-10 Bronco observation airplane.

1994-2002: Nonprofit Education/PAT Projects: by 1995, we moved into the Lancaster CA International Headquarters of the Society of Experimental Test Pilots (SETP). The NASA Dryden Flight Research Center (DFRC) awarded contracts to work with teachers in schools K-12 to develop lessons based on past and current flight research programs. Our staff grew to 10 people. We engaged test pilots and NASA engineers to work with teachers and gave them tours at DFRC. In 2002, I worked with the President of the SETP, Jimmy Doolittle III, to raise $50 million in funding for a new education center, museum, and SETP headquarters to be developed at the border between Palmdale and Lancaster on Avenue M.

The 2000s:

On September 29, 2007, Neil Armstrong lectured on the LLRV and LLTV at a Disneyland hotel in Anaheim, CA. I provided him with photos and technical data in August. I took him to lunch after his speech and proposed to organize a conference to examine the options for the return to the moon Constellation program lunar landing training. Neil encouraged me to proceed, and though he could not attend March 2008 at the Go For Lunar Landing Conference, we held it with about 200 attendees from the industry and NASA Centers. He could attend a conference I had initiated at the NASA JSC in Houston on December 9, 2008. Hosted by the Altair Lunar Lander Project Office, Gene Cernan, John Young, Harrison (Jack) Schmitt also participated.

The 2010s:

In 2012 I attended a Spacefest conference in Tucson, and again in 2017, and participated on an X-15 Panel with Joe Engle: Apollo astronaut and Commander of the second Orbital Space Shuttle Mission. Of twelve X-15 pilots, Joe is the only X-15 pilot still living today.

In 2016 I attended a conference to give a paper on Flying Spacecraft in Houston at the National Space Biomedical Research Institute. I met Angus Rupert there, and his friendship has matured. He is a neurologist with a Ph.D. in aerospace and headed US Army Aeromedical Research Laboratory for 12 years. His work includes Spatial Disorientation, Accident Investigation, Multisensory Cueing, Vestibular Psychophysics, Balance Prostheses. He flies his 1980 Cessna 185 tail dragger and attends the frequent Antique airplane Fly-Ins from his private hangar in Pensacola.

The 2020s

On January 7, 2020, in Orlando, I joined a 2-hour Engineering Apollo Panel discussion at The AIAA Scitech **2020** national conference. It was moderated by John Tylco (MIT). Frank Hughes represented Apollo ground simulators; I represented Apollo free-fight simulation and astronaut Dave Scott (Apollo 15 Commander); pictures are primarily contained in three of the five appendices, about half of the approximately 200 pictures.

On November 4, 2021, at the San Diego Bahia Resort Hotel, I presented a paper at the American Control and Guidance Systems Committee(ACGSC) on the Lunar Landing Research Vehicle (LLRV) Fly-By-Wire (FBW) History, the first aircraft to use FBW on October 30, 1964.

There were many events and people advancing flight in the 1920s, and I was fortunate throughout my career to encounter many of those people although I was not born until 1933. Although I worked with Jimmy Doolittle III, but never met his grandfather, I did meet or work the other six mentioned:

The first story of Jimmy Doolittle born in 1898, gives the reader a perspective of the genius of Doolittle as he ably deserved his medals of honor and status a national hero.

My boss, General Schweitzer, (1906 - 1996), was flying in 1929, and helping Chiang Kai Shek in 1932 to 1935 start his Air Force in China. He worked as a pilot for American Airlines 1936.

Hanna Reitsch who was flying gliders as a teenager in the early 1920s. She pioneered as a woman German test pilot in the 1930s and 1940s for helicopters, rocket airplanes, and the V-1 buzz. bomb.

M. G. Bekker, (1905- 1989), a Polish Engineer. an expert in tracked vehicles (tanks), after the invasion of Poland in 1939, had to move many times until he joined General Motors in Santa Barbara and was responsible for the design of the Apollo Lunar Rover wheel design.

Richard E. Day, (1917 -2004

Betty Love, 1922 turned 100 July, 22 2022. Betty and I started our careers in 1955, her first job title was "computer". as she was using a Frieden calculator.

Whitey Whiteside, born late 1920s, died late 1990s. He Joined the Army Air Corp at 16 years old with fake ID, rose to rank of Col. in charge of aircraft maintenance at Edwards AFB.

Jimmy Doolittle
(Rare pictures in Appendix A)

Doolittle died on Sept. 27, 1993, at age 96. Survived by his two sons, James, Jr. and John. Both men followed in their dad's footsteps by becoming Air Force officers. I worked with the grandson, Jimmy Doolittle III, the President of the Society of Experimental Test Pilots in 2002.

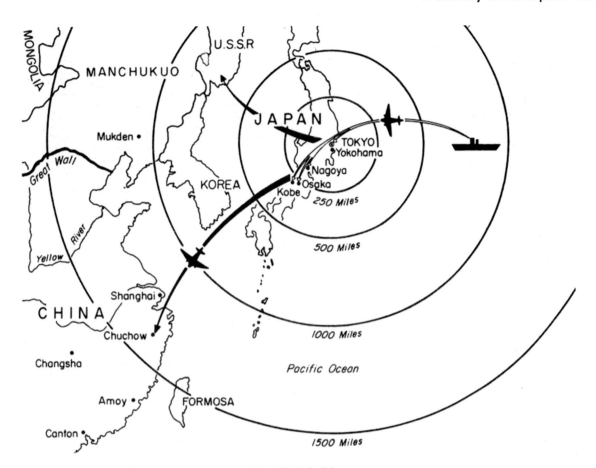

Doolittle's Plan

May, 11, 1942
President Roosevelt Awards the Medal of Honor to Brigadier General Jimmy Doolittle

In 1985, at 88:
Doolittle was promoted to full General by President Regan & Barry Goldwater

From: https://www.defense.gov/News/Feature-Stories/story/Article/2998360/
medal-of-honor-monday-army-air-corps-gen-jimmy-doolittle

Robert A. Hoover

(Photos in Appendix A The following research excerpts are from *Forever Flying R. A. "Bob" Hoover ISBN 978061537616*

Robert A. Hoover was born in Nashville, Tennessee, on Jan. 24, 1922. In his book, he remembered his excitement as a 5 - year old as he learned about Charles Lindberg's 1927 solo flight across the Atlantic. At 13, during his early experience as a paper boy finding a Model A chassis, he paid $7 for it and got it running so he could drive it. At 15, his early heroes, Roscoe Turner, Jimmy Doolittle, and Eddie Rickenbacker, motivated him to work 16-hour days sacking groceries to earn only $2, enough for 15 minutes of flying lessons. It took most of the year for him to overcome chronic motion sickness. Finally, at 16 years old, he soloed, as I later did in 1949. Soon after, he performed for his family, flying a 40 HP Piper Cub between two trees and under overlapping branches.

Bob Hoover's book tells his WWII combat flying, POW experience, and Post War test pilot accomplishments culminating in his last air show performing acrobatics at 85 years old.

Brigadier General John M. Schweizer:

My boss in 1969 and my stories are in in the 1960.s

Excerpts from: https://www.af.mil/AboutUs/Biographies/Display/Article/105688/brigadier-general-john-m-schweizer-jr/

John Mel Schweizer Jr. was born in Los Angeles, Calif., in 1906. He attended Occidental College there and graduated in 1929. He entered Army Air Corps Primary Flying School, March Field, Calif., in October 1929, graduated from advanced Flying School at Kelly Field, Texas, in October 1930, and was commissioned a second lieutenant in the Air Reserve. He spent the next two years on active duty with the Third Attack Group at Galveston, Texas.

From 1932 to 1935 General Schweizer was an aviation advisor and flying instructor for the National Government of the Republic China, as a member of an American group that went to China to organize and train an air force for Generalissimo Chiang Kai Shek.

Returning to the United States in the summer of 1935, he joined the 93rd Observation Squadron at Crissy Field, Calif., remaining until the spring of 1936 when he went on inactive duty and joined American Air Lines as a pilot. From January 1937 to May 1942 he was employed by the Humble Oil and Refining Company of Houston, Texas -- first as chief pilot and then as manager of aviation sales.

Recalled to active duty as a captain in the Air Force Reserve, General Schweizer served between September 1942 and January 1944 as a group commander in the Advanced Twin Engine Flying School at Ellington Field, Texas, attended the B-24 Transition School at Smyrna, Tenn., then was named director of flying and subsequently director of training at the B-24 Transition School at Tarrant Field, Fort Worth, Texas.

He served as deputy assistant chief of staff, A-3, Army Air Forces Central Flying Training Command, Randolph Field, Texas, from January 1944 to April 1945, when he was reassigned to the Sixth Air Force at Albrook Field, Canal Zone as deputy chief of staff and acting chief of staff. In June 1946 he returned to his former position with the Humble Oil and Refining Company.

General Schweizer accepted a regular commission in the Air Corps in July 1946 and served as commanding officer of Waller Air Force Base, Trinidad, British West Indies, until April 1948, when he was assigned to U.S. Air Force headquarters, Washington, D.C., as executive to the director of intelligence. In 1950 he attended the Industrial College of the Armed Forces at Fort McNair, Washington, D.C., and after graduation he was assigned to Tactical Air Command headquarters as deputy chief of staff for intelligence.

He was appointed assistant chief of staff for intelligence, U.S. Air Forces in Europe, with headquarters in Wiesbaden, Germany, in January 1953. General Schweizer was named assistant chief of staff for intelligence, Supreme Headquarters, Allied Powers Europe at Paris, France, on May 10, 1954. He is rated a command pilot.

From: https://www.latimes.com/archives/la-xpm-1996-01-26-mn-29063-story.html

LA Times, OBIT: John M. Schweizer Jr.; "Retired Air Force Brigadier General, Executive, January 26, 1996"

John Mel Schweizer, Jr., 89, retired Air Force brigadier general who moved from pilot to executive in military intelligence and then in commercial aviation. A native of Los Angeles, Schweizer studied at Occidental College, where he was quarterback of the football team and lead trumpet for the Earl Bernett Orchestra. As a pilot in the Army Air Corps in the 1930s, he trained Nationalist Chinese military pilots (1932 as a 1ˢᵗ Lt he told me.). He worked as a commercial pilot for American Airlines and Humble Oil until he was recalled to active duty in World War II. After becoming a brigadier general in 1953, Schweizer served as Air Force director of intelligence in Europe and later as director of intelligence for the Allied powers in Europe. He was honored for his service by the United States, France, Italy and Greece. Schweizer retired from the Air Force in 1959 to become the European representative for Bell Aerosystems Corp. in Paris and later Bell vice president in Los Angeles. In 1971, he became executive director of the Voices in Vital America supporting the recovery of MIAs and POWs in Vietnam. [He died] on Monday in Camarillo of lung cancer on Monday in Camarillo of lung cancer

Hanna Reitsch

(My story is in **1970s Decade,** I have a link to her auto biography in pdf included in Appendix E, Many photos of her are also included in Appendix A SETP Photos From *"History of the First Twenty Years".*

The following excerpts of information has been found by my research on Wikipedia: https://en.wikipedia.org/wiki/Hanna_Reitsch

Hanna Reitsch was born March 29, 1912 in Hirschberg a historic city in Southwestern Poland. She died Aust 24, 1979 (age 67) in Frankfurt am Main, Hesse, West Germany. While a medical student in Berlin in1932,she began flight training at a School of gliding. In 1935 she became a test pilot testing dive brakes for gliders and troop carrying gliders. Adolf Hitler awards Hanna Reitsch the Iron Cross 2ⁿᵈ Class in March 1941 Reitsch was the first female helicopter pilot and one of the few pilots to fly the first fully controllable helicopter inside the Deutschlandhalle for three weeks for which she received the Military Flying Medal in 1938. In September 1938, Reitsch flew the DFS Habicht in the Cleveland National Air Races. Reitsch was a test pilot on the Junkers JU 87 dive bomber and Dornier Do-17 light/fast bomber projects, for which she received the Iron Cross, Second Class, from Hitler on 28 March 1941. Reitsch flew the rocket-propelled Merrerschmitt Me 163 Komet in 1942. A crash landing on her fifth Me 163 flight badly injured Reitsch; she spent five months in a hospital recovering. Reitsch received the Iron Cross First Class following the accident, one of only three women to do so. On 28 February 1944, she presented the idea of Operation Suicide to Hitler., which *"would require men who were ready to sacrifice themselves in the conviction that only by this means could their country be saved."* Although Hitler *"did not consider the war situation sufficiently serious to warrant them ... and ... this was not the right psychological moment",* he gave his approval. The project was assigned to Gen Gunther Korten. There were about seventy volunteers who enrolled in the Suicide Group as pilots for the human glider-bomb. By April 1944, Reitsch and Heinz Kensche finished tests of the Me 328. The plan was never implemented operationally. Former British test pilot and Royal Navy Officer Eric Brown said he received a letter from Reitsch in early August 1979 in which she said, *"It began in the bunker, there it shall end."* Within weeks she was dead. Brown speculated that Reitsch had taken the cyanide capsule Hitler had given her in the bunker, and that she had taken it as part of a suicide pact with Greim. No autopsy was performed, or at least no such report is available.

Hanna Reitsch E-Book *"Flying Is MY Life"*

https://www.barnesandnoble.com/w/flying-is-my-life-illustrated-edition-hanna-reitsch/1124240790

M.G. Bekker
(The Father of Off-Road Mobility) My Story in the 1970s Decade

Obituary: *Mieczysław Gregory Bekker (1905–1989) was a Polish engineer and scientist.*

Bekker was born in Strzyżów, near Hrubieszow, Poland, and graduated from Warsaw Technical University in 1929. Bekker worked for the Polish Ministry of Military Affairs (1931–1939) at the Army Research Institute (Wojskowy Instytut Badań Inżynierii) in Warsaw. There he worked on systems for tracked vehicles to work on uneven ground. In the Invasion of Poland, he was in a unit that retreated to Romania, then moved to France in 1939. In 1942 he accepted the offer of the Canadian government to move to Ottawa to work in armored vehicle research. He entered the Canadian Army. In 1943 as a researcher, and reached the rank of Lieutenant Colonel. Decommissioned in 1956, he moved to the U. S. He was assistant professor at the University of Michigan and worked in the Army Vehicle Laboratory in Detroit. In 1961 he joined General Motors to work on the lunar vehicle project. He was a leading specialist in the theory and design of military and off-the-road locomotion vehicles and an originator of a new engineering discipline called terramechanics ". Bekker co-authored the general idea and contributed significantly to the design and construction of the Lunar Roving Vehicle used by missions Apollo 15, Apollo 16, and Apollo 17 on the Moon. In addition, he was the author of several patented inventions in the area of off-the-road vehicles, including those for extraterrestrial use. He wrote many papers and articles and the book "Theory of Land Locomotion." Bekker died in Santa Barbara on January 8, 1989.

Richard E. Day
1917 - 2004 My Story is in the 1960s Decade

Joe Fletcher My story in the 1970s Decade: My Landlord in Santa Monica Canyon 1970-1975, The last four pages of Appendix E describe his story of landing on the North Pole.

My research has found the following information on Wikipedia: dia.org/wiki/Joseph_O._Fletcher

Joe Fletcher was born on 5/16/1920 in Ryegat, Montana. He died on July 6, 2008, in Sequim, Washington, at age 88.

Fletcher started studying at the University of Oklahoma. He then continued his studies in meteorology at the Massachusetts Institute of Technology. After graduation, he entered the U.S. Army Air Corps and became the deputy commanding officer of the 4th Weather Group, United States Air Force, stationed in Alaska.

On March 19, 1952, his team landed with a C-47 aircraft, modified to have both wheels and skis, on a tabular iceberg in the Arctic Ocean and established a weather station there, which remained manned for 22 years before the iceberg broke up. The station was initially known as *"T-3"* but was soon renamed *"Fletcher's Ice Island."*

On May 3, 1952, pilot William P. Benedict and Fletcher, as co-pilot, flew that plane to the North Pole, becoming the first humans to land there and the first humans (together with scientist Albert P. Crary, (who flew with them) to set foot on the exact geographical North Pole. (However, some sources credit this achievement instead to a Soviet Union expedition that landed there on April 23, 1948.)

Fletcher left the Air Force in 1963. In later years, he held various management positions in meteorological institutions, including a post as director of the NOAA's Office of Oceanic and Atmospheric Research (OAR). Fletcher received a doctorate from the University of Alaska in 1979. He retired in 1993. In 2005, he was awarded the honorary membership of the American Meteorological Society.

history.com (research source)

A ski-modified U.S. Air Force C-47 piloted by Lieutenant Colonel Joseph O. Fletcher and Lieutenant Colonel William P. Benedict became the first to land on the North Pole. A moment later, Fletcher climbed out of the plane and walked to the exact geographic North Pole, probably the first person in history to do so.

In the early 20th century, American explorers Robert Peary and Dr. Frederick Cook, both claiming to have separately reached the North Pole by land, publicly disputed each other's claims. Finally, in 1911, Congress formally recognized Peary's claim. In recent years, further studies of the conflicting claims suggest that neither expedition reached the exact North Pole but that Peary came far closer, failing perhaps 30 miles short. In 1952, Lieutenant Colonel Fletcher was the first to stand on the North Pole undisputedly. Standing alongside Fletcher on the top of the world was Dr. Albert P. Crary, a scientist who, in 1961, traveled to the South Pole by motorized vehicle, becoming the first person in history to have stood on both poles.

Betty Love

Born July 8, 1922, my story in the 1990s Decade https://www.nasa.gov/50th/50th_magazine/index.html People/ NAS. Innovators and Unsung Hero's

NASA's Innovators and Unsung Heroes

NASA's 50-year legacy of pioneering the future is as much the product of quietly persistent innovators and unsung heroes as it is of the agency's more high profile representatives. The following profiles represent just a sample of people who have made a difference in the agency's history.

Betty Love: A living link to the early X-planes
By Christian Gelzer

Betty Love came to work for the National Advisory Committee for Aeronautics (NACA) High Speed Flight Research Station at Muroc Air Force Base, now Edwards Air Force Base, Calif., in 1952. In a sense, she has never left.

Unlike most of the employees at the NACA station at the time, Betty was a native Californian. She grew up not far from what is today NASA's Dryden Flight Research Center, and can remember family bonfires on Rosamond Dry Lake and trips by car across Rogers Dry Lake at night in 1942.

Love began working as a "human computer" barely six years after the first contingent of NACA engineers and technicians arrived at the remote desert base. Her job was to take film from an X-plane after a research flight, time code the traces on that film, turn that data into numbers and graph the numbers so the engineers could see what the strain gages registered or how much the control surfaces had deflected.

Honors for X-15 work - X-15 research engineering technician Betty Love (on right) joined her husband, X-15 project manager Jim Love and test pilot Bill Dana for a ceremony honoring the program's success in California Gov. Ronald Reagan's office (1969). From left, state assemblyman Kent Stacey, Judi Dana, Bill Dana, Gov. Ronald Reagan, Jim Love, Betty Love and state senator Walter Stiern.

She started as a GS-1 – the lowest rung on the government pay scale – and recalled that "It was several years before my two-week paycheck would equal $100."

Working conditions were not what most people expected, although Love was not surprised. "The wind blew in the winter, and it was cold and dusty," she recalled. "The wind blew in the summer, and it was hot and dusty." Each morning as the human computers came to work – they were all women – they used their government-issued dust broom to sweep the dust and sand off their desks. "It was like a library," she recalled.

"No one spoke to anyone [except] in hushed tones. We were allowed to get coffee twice a day. The engineers had a lot more fun in their offices."

Love's work gradually began to change. Instead of just reading traces, she began to receive special instructions with her film: look for amplitude, or count the frequencies. In 1954 she was taken out of the computers' office and put in a room with four engineers who worked on structures.

"I still had my Friden calculator," she added, but she began reducing data just for those engineers. Moreover, unlike her previous workplace, "I could ask what I was doing, what it was going to be used for, and I did ask lots of 'whys.'" No longer just a computer, Love was promoted to aeronautical research engineering technician. Along with the title came new responsibilities. "I worked on all the research planes from the X-1 and the D-558-II to the X-15 and the XB-70," she

"I still had my Friden calculator," she added, but she began reducing data just for those engineers. Moreover, unlike her previous workplace, "I could ask what I was doing, what it was going to be used for, and I did ask lots of 'whys.'" No longer just a computer, Love was promoted to aeronautical research engineering technician. Along with the title came new responsibilities. "I worked on all the research planes from the X-1 and the D-558-II to the X-15 and the XB-70," she remembered.

When telereaders came along, her work became somewhat less laborious. She could now mark film traces with a foot pedal while entering the data on a keyboard, instead of writing the numbers by hand in columns. The cards were then loaded into a sorter that generated the graphs. Before long, Love was writing short computer programs to reduce the data more directly. Not surprisingly, she was given the first programmable desk computer the center purchased. Later in her career with NASA, Love co-authored several technical papers and served as third author on others. She never found her years at the center tedious or unrewarding, and it was often hard to leave work on time.

"The car pool used to give me fits," recalled Love of the early years, since she was invariably the last to reach the car, and the others in the pool were eager to get home. Until, that is, "Neil Armstrong joined the car pool; then they left me alone and got all over him for not being ready at four o'clock."

Love technically retired in 1973 but continues to be invaluable as a volunteer in Dryden's history office – a living, on-going link to the earliest years at the center, smoothly bridging the NACA and NASA periods.

Honors for X-15 work - X-15 research engineering technician Betty Love (on the right) joined her husband, X-15 project manager Jim Love and test pilot Bill Dana for a ceremony honoring the program's success in California Gov. Ronald Reagan's office (1969). From left, state assemblyman Kent Stacey, Judi Dana, Bill Dana, Gov. Ronald Reagan, Jim Love, Betty Love, and state senator Walter Stiern.

NASA's Innovators and Unsung Heroes

Betty Love: A living link to the Early X-planes

By Christian Gelzer

Betty Love came to work for the National Advisory Committee for Aeronautics (NACA) High Speed Flight Research Station at Muroc Air Force Base, now Edwards Air Force Base, Calif, in 1952. In a sense, she has never left.

Unlike most of the employees at the NACA station at the time, Betty was a native Californian. She grew up not far from what is today NASA's Dryden Flight Research Center and can remember family bonfires on Rosamond Dry Lake and trips by car across Rogers Dry Lake at night in 1942.

Love began working as a *"human computer"* barely six years after the first contingent of NACA engineers and technicians arrived at the remote desert base. Her job was to take film from an X-plane after a research flight, and time code the traces on that film, turn that data into numbers, and graph the numbers so the engineers could see what the strain gages registered or how much the control surfaces had deflected.

She started as a GS-1 - the lowest rung on the government pay scale. Love recalled it was several years before her two-week paycheck would equal $100. Working conditions were not what most people expected, although Love was not surprised. "The wind blew in the winter, and it was cold and dusty," she recalled. "The wind blew in the summer, hot and dusty." Each morning as the human computers came to work, they were all women. They used their government-issued dust broom to sweep the dust and sand off their desks. "It was like a library," she recalled.

"No one spoke to anyone [except] in hushed tones. We were allowed to get coffee twice a day. The engineers had a lot more fun in their offices." When telereaders came along, her work became somewhat less laborious. She could now mark film traces with a foot pedal while entering the data on a keyboard instead of writing the numbers by hand in columns. The cards were then loaded into a sorter that generated the graphs. Before long, Love was writing short computer programs to reduce the data more directly.

Not surprisingly, she was given the first programmable desk computer the center purchased. Later in her career with NASA, Love co-authored several technical papers and served as a third author on others. She never found her years at the center tedious or unrewarding, and it was often hard to leave work on time.

"The carpool used to give me fits," recalled Love of the early years since she was invariably the last to reach the car, and the others in the pool were eager to get home. Until that is, "Neil Armstrong joined the carpool; then they left me alone and got all over him for not being ready at four o'clock." Love technically retired in 1973 but continues to be invaluable as a volunteer in Dryden's history office, a living, persistent link to the earliest years at the center, smoothly bridging the NACA and NASA periods.

Wayne and Betty Love (on my PAT Projects Board), circa 1997

Walter (Whitey) Whiteside

Born in the late 1920s, my story is set in the 1990s Decade.

1930's Decade

I was born October 22, 1933 in Phoenix Arizona

1940's Decade

Ed Rhodes, the Bell project engineer in 1942 for the first American jet airplane, when ready for retirement in 1976, was assigned to be my administrative assistant to help me manage a large (120) group of engineers and technicians for 18 months delivering three Lunar Landing Training Vehicles.

The First US Jet, the XP 59

Working With Mac

I worked on the J79 with Mac (Angus MacEachern) from GE Service Engineering at the Air Force Power Plant Branch in 1956. He told me a story of riding a military train, closely guarded by U. S. Army troops in September 1942, from the Bell Aircraft plant in Buffalo to Muroc Army Air Field (now Edwards AFB). Mac was accompanying two W. 1's British jet engines. Mac said he rotated the two compressors every few hours (his on-duty chore as a crew member per the next two pages) to keep the bearings from taking a set. These engines were installed in the Bell XP-59A Airacomet aircraft in Buffalo, NY and the book shown below tells the story on page 28 of that book.

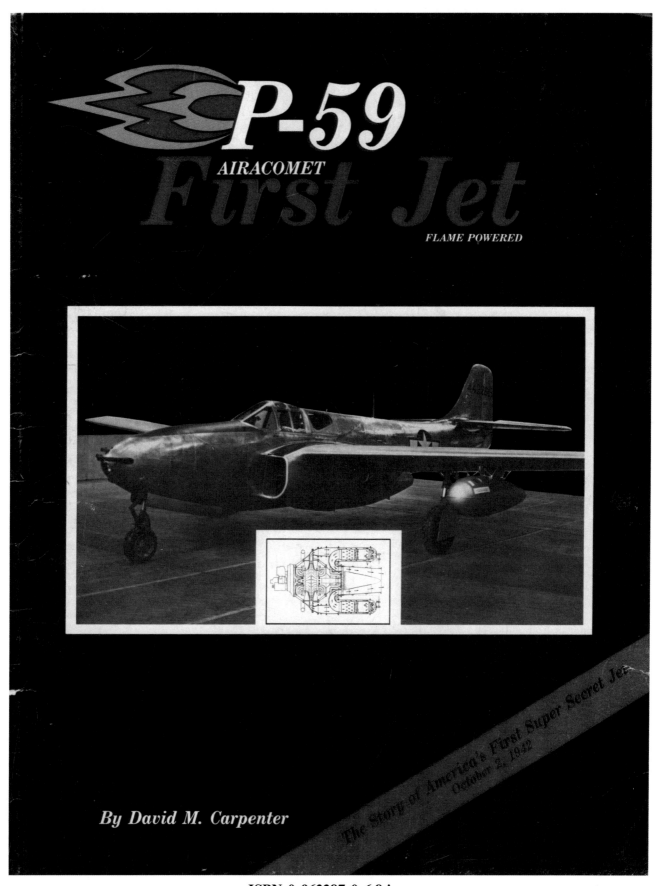

Flying: the XP-59A on rails
as told by Ted Rogers, a member of the GE Field Team

Early August saw engines, called Type I Superchargers for security reasons, being loaded into a boxcar under armed guard at Boston's South Station headed for Bell Aircraft, Buffalo, N.Y. Frank Burnham and I, selected to accompany the engines and supervise their installation in the airplane, spent the evening thoroughly briefing ourselves on the aircraft manual. It was the first time we had seen it. The airplane itself, when we saw it next morning at the Main Street plant of Bell, was a mid-wing monoplane with wing loading so low it looked like a powered glider. It weighed about 9,000 pounds, had a span of 49 feet and a total length of 38 feet. The tricycle gear, built for jet engines, allowed it to sit low and level, giving excel lent ground visibility. The tail section curved high to ensure clearing the jet wake. August was spent fitting the engines in the airplane, re-routing fuel and air lines, rigging controls and completing the final assembly of the aircraft.

In early September, it was decided to ship the engines mounted in the plane to avoid delays of removal and re-installation. A stumbling block - a fear of damaging the bearings - we thought could be overcome by rotating the engines slowly while in transit. Some body, of course, had to go with the plane to keep these engines rolling. We were determined to protect them at all costs, so with **Angus MacEachern**, Lieut. Woolford, a Sergeant and two Privates from Wright Field added to our crew, we prepared for the "flight. Midnight, September 12, 1942, at the yards of the Niagara Power Company in Buffalo, Air Force guards patrolled the deserted area while a huge overhead crane:: lifted its precious cargo and set it down gently on the railroad flat cars. One crate contained the fuselage; another held the two wings. With previously prepared blocks, a gasoline powered air compressor was secured to the rear half of the flat car containing the fuselage, while a coil of copper tubing was unwound and connected in the interior of the crate. One could feel the unspoken questions of the yard crew - what could possibly be in the crate to use compressed air? Some new explosive? A compressed air driven airplane? It was just *"Queenie."* Actually, *"Queenie"* was at this particular time nameless. It was not until we had pulled the cowlings off several hundred times that some joker nicknamed her *"Queenie"* for the lady of burlesque fame.

We hastily completed coupling the air line and tried our system of propulsion. I was scared! I had proposed this method of rotating the engine. Herc we were all set to take off for California. What if it didn't work? With a supreme show of self-confidence, we filled the gasoline tank, checked the:: engines for freeness and started the compressor. Meanwhile, the crates had been secured, and a safety rail placed around the open end of the car. We opened the valve feeding the air jets to the turbines- and the engines rolled! Climbing into the cockpit, I checked the engine speed at 400 rpm and observed that oil was flowing protectively to the bearings. The two flat cars and a combination mail and passenger cars for our home enroute were coupled together, and with both engines purring we were off, coupled to the "Red Ball," California-bound fast freight.

Our "home" which we slowly put in order, was Vintage 1900. In the mail section we had six drums of gasoline, safety cans, filters and tools, plus an old ice chest stocked with what we considered the necessities of a "flight" to the coast for three persons. Someone had neglected our Army Air Corps companions. The passenger section of the car had several seats, on which we placed mattresses to make bunks. Our kitchen consisted of a Sterno stove and a case of canned heat. The salesman who sold us that stove has probably since made a fortune in Alaska selling refrigerators. It was just not intended to cook for seven and we discarded it after it proved unstable in *"flight"* - twice spilling our food the length of the car. The rest of the night was spent in securing the various pieces of equipment, as the jolting and swaying of the train proved necessary. It's no fun to hold down a one-ton compressor while fastening its ropes on a flat car traveling 60 mph or pouring gasoline into its ever-hungry mouth, while the cars hammer around a curve.

We lived through the first night, but mid-morning of next day the forward flat car bearings burned out. With incredible speed, a railway crane appeared, and within two hours the entire assembly had been changed. The rest of the day passed uneventfully until around 1:00 a.m., when I was awakened by loud voices and persistent shaking. Adjusting my eyes to the sparse light of the lantern, I saw uniforms and guns. Was this what they had warned me about - saboteurs! But no - more light and the bull-like voice of a Marine Sergeant convinced me that the trouble was not with saboteurs. A car of high explosive had been coupled to the front of our "home" and a squad of Marines accompanying it insisted we were in their accommodations. We compromised by letting them bed down in one end of the car while an armed truce prevailed.

After a sleepless night, we pulled into Kansas City where the Marines and most of our crew left during the time the car of explosives was shunted around. If you want a thrill ride down a railroad "hump" behind a car labeled in large red letters "High Explosives - Do Not Hump." We stood on the fuselage car while this was accomplished with a teeth-loosening jar. We soon convinced the railroad that we wanted no more of that, and the car with the Marines was relegated to the end of the train.

We did take advantage of the stop to run, in turns, to the nearest restaurant for a hot meal. You can imagine our surprise, Burnham's and mine, upon returning to find our car nowhere in sight!

A hectic search of the yards, assisted by a yard policeman who was convinced that we belonged in a nut house, finally located it. We were determined not to try that again. Enroute just about dusk after we had filled the compressor gas tank from a new drum, without warning it died. While the air hissed inexorably through the lines, we frantically tried to restart the compressor - useless. The ignition checked okay, so it must be the fuel. By frantic manipulation of the choke, it restarted just as the jet engines were coming to rest. The air pressure slowly climbed back up as we sighed with relief - then the engine quit again. This time we saved the day by blowing out the sediment bulb. This was only the beginning. Even though we replaced the dirty gas with that from another drum, we spent the whole night cleaning sediment and manipulating the choke and mixture needles during the off cycle of the compressor. We finally succeeded in removing all the dirt but the fuel system gave us frequent alarms all the rest of the trip.

Tuesday and Wednesday we spent catching up on our sleep in relays, eating sandwiches fouled by gasoline and washing them down with sips of warm water. Twice we were parked in yards next to a carload of sheep! The food bill was low that day.

Thursday found us still plodding across country - engines purring. We stopped at Clovis, N.M., for water. Upon assurance the train would stand there for 20 minutes, we snatched a hot meal and raced back to it - all but Frank. He miscalculated, and we pulled out without him. Some miles further we swung into a siding to repair a car. A passenger train slowed as it passed, and Frank jumped off to rejoin us. Three hours in this barren land convinced us that we were the "foul ball" express. A protest wired to Wright Field convinced the railroad to cut off the car needing repairs and get us rolling.

On Friday we really set a pace across Arizona, halting only long enough to refuel and replenish water for the locomotive. Up through the mountains, cold and capped with snow, we climbed; then down into a green valley with palm trees lining the streets. We were in San Bernardino. Here we were switched from the main line and headed for Muroc. We passed our first peaceful night and early morning found us traveling across an unbroken vista of plain nothing, broken only by an occasional Joshua tree. Then Muroc! Off in the distance we could see the reflected sheen of a huge lake. The brakeman dispelled our impression. There was no lake; it was a mirage. Moving to a siding, we cut the air compressor. We were saddle sore and filthy as we climbed down from the first coast-to-coast *"flight"* of a jet airplane - six full days.

My Connections to the 1942 Crew

In October 1942, at Muroc Dry Lake, CA, the two I-A engines powered the historic first flight of the first jet-powered aircraft in the United States. The story by David M. Carpenter in the publication P-59 Air Comet First Jet Flight on October 2, 1942. August 27, 1939 the German Heinkel He 178 made its first flight.

Years later (1966), Ed Rhodes, told me a story of Mac sitting in the cockpit of the first Bell XP-59A, which had the newly designated J31 engines installed. First, Mac would pressure check the cockpit by smoking cigars inside, and then Ed would look for the leaking cigar smoke outside.

Mac was a prankster: One of the Air Force technicians was making a hardwood toolbox with complex precision joints, and when he left the parts on the bench to heat the glue for assembly, Mac would quickly cut a little off one of the matching parts, and his victim would wrestle with the misfit, having to re-make the parts again. Mac got caught on the next try or two, and the victim promised Mac would never know when the retaliation would come. Sure enough, months later, Mac was in San Bernardino on the weekend and got a call from the Lancaster Fire Chief: *"Mac, you better get home as your house is on fire, and we're doing what we can to save it."* So Mac rushed home, and when he was about a block away from it, he thought that this might be retribution.

A Seven-minute video detailed story of Bell's test pilot Bob Stanley's first taxi tests and first unofficial flight October 1, 1942 and official first flight October 2, 1942.

https://youtu.be/joq0qb4OI9o

The video above confirms Larry Bell born in 1894, founded Bell Aircraft in 1935).also occupied the second seat, and instrumentation recorders were added in a compartment aft of the cockpit for later flights.

Ed Rhodes, Bell XP59 Project Engineer

<u>Ed Rhodes, Added Flight Engineer's Seat, Just in front of Cockpit--- In 1966 my Bell administrator for 18 months as I served as Technical Director building three Lunar Landing Training Vehicles</u>

The GE crew: Frank Burnham, Ed Tritle, Roy Shoults, Ted Rogers and Angus McEachern, October 2, 1942. (Photo courtesy of J. Brown)

Bell's design team: H. Poyer, R. Wolf, E. Rhodes, Jim Limage and H. Bowers, five of the "Secret 6", before the first flight - Brian Sparks is missing. (Photo courtesy of R. Wolf)

Mac & Ed's Crew

Whittle 1.A at GE Lynn, MA River Works Plant

Moving the second XP-59A to a new site due to flooding of "Dry" Lake. Note: fake propeller made by Joe Brown standing second from left and Lt. B. Van Doren in foreground. (Photo courtesy of J. Brown)

The Wartime Secrecy at Work

1949 My High School Flying Club

WWII Pilot Training from the Phoenix Sky Harbor Airport had a dedicated hangar facility to train Chinese pilots. Our Phoenix Union School District, including the Link Trainers, was awarded those facilities. There was a fleet of about twelve 65-hp Aeronca Champions as primary trainers, with a Stinson Voyager and a Cessna 150 completing the aircraft inventory. My first instructional flight was on November 10, 1949, shortly after my 16th birthday. The total cost was $172.50 at just $3/hour for gas and oil. Instructors were not charged as their salary was paid by the school district and included all the ground school instruction. I earned my private license on May 26, 1951, at age 17.

One of my solo practice flights over the Indian reservation east of Scottsdale was undoubtedly memorable. On the way out to the practice area, I noticed a B-25 WWII bomber entering the airspace, and I was impressed with its much greater speed than my Aeronca. I was flying over the practice area a massive dust storm formed over the western part of the valley, and the rush was on by the many light airplanes in the airspace to get back to the airport and land before the storm reached the airport. The B-25 was on final approach, and the control tower, not realizing on my base leg I was too close, called me on my radio receiver to abort. My only choice was to make a full power climbing left turn, making the B-25 slipstream turbulence a ride, not by choice. I passed the ground school courses the following year for a commercial pilot's license but never completed the flight hours required to receive the commercial license.

1950's Decade

I received my Bachelor of Science Mechanical Engineering Degree (BSME) and served as a GE jet engine test engineer in 1955. After moving to California, GE Promoted me to a jet engine flight test engineer at Edwards AFB 1956 - 1960.

Among many, the shortest summer job I ever had lasted about six hours. It was to load three Stearman crop dusters out of Sky Harbor Airport in Phoenix in the summer of 1950. Reporting for work before dawn, told by an older teenager we would be riding in the front cockpit of the Stearman biplane to ride out to the loading site south of Chandler. Unfortunately, we flipped a coin, and I lost, so for the outbound flight, I sat on the paddle wheel blade at the bottom of the cockpit, which discharged the sulfur dust for crop spraying. The older teen sat on top of me.

When we landed on the dirt strip, hundreds of 60-lb. bags of sulfur dust were stacked along the fence. None of the Stearman's ever shut off their engines the whole time (two or three hours) and loaded by three loaders: One pulling the sack from the fence pile, one (me) tossing the bag to the loader on the wing, who then dumped it into the front cockpit hopper. After that, the pilot would take off, the next would land, and so on all morning.

The combination of the dirt from the landing strip and sulfur dust blowing directly from the propeller slipstream almost did me in, as my helmet was two sizes too big, and with every breath, I inhaled the mixture of sweat, sulfur, and dust. But, finally, I got to ride on top, going back to Sky Harbor, and could see a paved runway and an unused Williams AFB auxiliary field across the fence. Back in the office, it took me five minutes to quit.

Rewarded by the BSME

Several companies interviewed me before graduating in May 1955 from the University of Arizona. One offered their company-paid trip for interviews in New York City by Babcock and Wilcox, a major college engineering recruiting company. Several married couples with children were in the Mechanical Engineering program, and Hutch Davis was one year ahead of me. Hutch was a supervisor at the General Electric Engineering Training Program's Gas Turbine Test Department in Schenectady, NY. Hutch arranged for GE to interview me and pay for my travel by train to Schenectady from New York and back to NYC. Given GE's aerospace product lines, I accepted their offer. We were very grateful for Hutch and his wife Jo's hospitality and assistance in getting relocated. We moved into a small house in Scotia, NY, across the Mohawk River Bridge and about two miles from the main gate at the GE Schenectady historic main plant where Thomas Edison and associates founded the company in 1892. A local schoolteacher rented us the house for the three summer months, which matched the three months of the engineering training program assignment.

I worked on the 5,000-hp gas turbine locomotive engines on the second shift test program, supervised by Hutch Davis. As we had the only TV in the group, about five of us on the second shift, all married, would gather at our house in Scotia after work, make homemade ice cream, and watch Steve Allen on the first Tonight Show. The next morning we'd usually meet at one of the three company-owned golf courses that paid our green fees.

One night, while testing a Bosch diesel pump on Bunker "C" fuel (the most contaminated grade) in the basement of an abandoned building, the discharge hose from the pump became unsecured on the wooden slat floor. It was flowing hot oil at about 180 degrees Fahrenheit and 30 psi. I tripped on the hose, then it came out of the tank and started peeling the skin on my back from my waist to my shoulders. Screaming, I pushed it off my waist and, as Hutch was standing next to me — right into his crotch. He immediately knocked the snaking hose spraying hot oil away from both of us. We both were jumping and yelling, slipping on the oiled slat floor, when a tech hit the master breaker on the wall, which turned off the pump, but the lights with it.

After slipping and yelling, the breaker turned on for lights. Hutch and I headed up the stairs to exit the building. Then we headed for the main gate for the infirmary. The techs had called for the ambulance. We were almost to the infirmary when it got to us. We ran about a quarter mile. The nurses took scissors and cut the tee shirt off my back, leaving my entire back embedded in the oil with what was left of the shirt. Next, they sprayed me with a foam medication, which worked very well. They never hospitalized me but took me home a few miles away — bound there for several days. Several times a day, they shuttled me back and forth as my back filled with large blisters; treating me well, they left no scars. Hutch did okay, too; we were both fortunate. My only follow-up was to get my graduation gift — a Longines watch — cleaned, as it was full of Bunker "C."

I was pitching fast softball for the GE Schenectady team, and after moving some heavy motors during work, I pitched a game in Pittsfield, MA, on a Saturday night. I came home to Scotia in great pain with an acute strain in the small of my back. Later that night, the company ambulance transported me to the hospital, where the treatment in 1955 was to hang 40 lbs. off each leg in traction, which took ten days to get the job done. My roommate was a teenager who'd shattered his leg in a baseball game. His parents smuggled my 18-month-old son, Allen, up the fire escape to see me, as the hospital would not allow him as a visitor. The Human Relations representative at GE would bring me a newspaper daily and eventually announce that they had won the appeal for a workman's compensation case based on lifting motors. I never used those benefits but I still have the back support girdle.

Before leaving upstate New York on the first assignment, Sondra and I took Allen to New York City on the weekend to visit my mother's first cousin Catherine Suder. Catherine, married to Charles Suder, Comptroller for Safeway Stores, in a high-rise apartment in New Jersey along the Hudson River. After retirement to Phoenix in the 1960s, we visited them with all four of our sons. Charles was full of funny stories and hailed from Prescott, AZ, where his father's name is inscribed on the plaque on the historic courthouse in the center of town.

Launch of my Aerospace Career

At the end of the three months in Schenectady, I accepted the next assignment at the GE Evendale Ohio plant (just north of Cincinnati), working on afterburners and performance evaluation of the J79 jet engine. Slated to be the primary powerplant in a modified Douglas F4-D used in our flight testing, we had to help perform a 50-hour PFRT (Preliminary Flight Rating Test) before it could make its first manned flight. Roy Pryor was the chief pilot who made the first J79 flight, and Dick Smith, whom I later worked for in 1957, was the chief engineer. In addition, an older engineer, Pender Love, helped me fit in on the rapidly moving program staffed round the clock, seven days a week.

I remember searching for hot turbine blades on the roof of the test cells when they separated from their turbine wheels and escaped to the roof through the exhaust stacks. There were banks of tall (maybe six or seven feet) mercury manometers on the walls of the cells to accurately measure the lower pressures and instrumentation, such as strain gauges on bearings; we often worked on assembly problems to accommodate this unique instrumentation. I was on second shift with lots of overtime and soon got rid of the '51 two-door Ford (which for two years was infamous for its occasional electrical shorts, leaving the engine dead and its passengers stranded). Subsequently, we bought a dealer demonstrator black and white, two-door 1955 Pontiac sport coupe; a natural uplift.

Arrival at Edwards Air Force Base
(EAFB), June 1956

I initially worked on the J79 engine for the field service-engineering department in offices in the old WWII barracks on the south base. My duties were to advise and support contractor and Air Force personnel in its operations and maintenance. Close by, GE Flight Test's offices and hangars housed the Douglas XF4D and the twin-engine McDonnell F-101 Voodoo, both modified as J79 test beds.

The GE J79 MACH 2 Engine (significantly lighter than the Pratt-Whitney competition

I was the only graduate engineer to work among veteran Korean War engine representatives on the F86 and F-100 fighters, supporting tests on other early Mach 2 aircraft. Lockheed F-104s, Convair B58s, Grumman F11F-1F Super Tiger, and Chance Vought Regulus II Missile all used the J79 engine.

The J79 was a significant advance in lightweight, high-thrust jet engines capable of Mach 2 flight. NACA (National Advisory Committee for Aeronautics), NASA's predecessor, was located on the south base until 1954 — then moved to the north base, across to the west side of Rogers Dry Lakebed. As the F-104 early flight test program progressed, the NACA High Speed Flight Station (later becoming the NASA Flight Research Center)--FRC at the north end of the main base received its first YF-104 (in August 1956).

Milt Thompson had been hired in early 1956 as a research engineer with hopes of becoming a member of the Test Pilot office. Instead, he had to wait till Scott Crossfield vacated the Chief Test Pilot position to join North American (who hired him to make the first contractor flights on the X-15). Milt's later rise to fly the X-15 and the early Lifting Bodies earned him national recognition. A humorous mention of me is in his book At the Edge of Space, which he autographed for me shortly before he died on August 6, 1993. His NASA friends had scheduled a roast for Milt in Lancaster, CA, the night of the day he died, and as the out-of-towners were all on the way, per his family's wishes, they roasted Milt anyway.

During the early F-104 flight tests, we GE engine reps would hang out in Fighter Ops, playing ping-pong with pilots waiting for their flights. One day, our game was interrupted by a control tower page for Captain James Woods and Captain Gordo Cooper. A mid-air collision occurred between a T-Bird on its climb after takeoff and a KC-135 tanker flying low over the Rogers Dry Lakebed. (Captain Woods retired as a Col. and prospective astronaut in the X-20 Dyna-Soar program. After discharge, he was the personal pilot for the Shah of Iran, flying his private Boeing 707 to escape the revolution in November 1979. Captain Cooper, three years later, became a NASA Mercury astronaut).

The T-Bird crashed, killing the Bell test pilot, and the KC-135, with two crewmembers, lost all electrical and hydraulic power. Jim Woods and Gordo flew chase in their T-Bird, escorting the crippled KC-135 to a safe wheels-up landing on the lakebed. The Air Force repaired the tanker at the base maintenance facility and returned for service.

Forty years later, I talked to Stan Butchart (the wingman to President H. W. Bush when he was shot down in the Pacific in World War II). Stan was the NASA research test pilot in the co-pilot's seat in that KC-135. Stan related how they got talked down for that landing: The chase T-Bird with James Wood and Gordo used a notepad and wrote cryptic notes, holding the notepad up to the side of their T-Bird canopy as they flew in tight beside the cockpit of the tanker. In other words, they communicated through the cockpit windows (probably done many times in emergencies) with written notes to learn the condition of tanker damage and get it down safely.

This conversation occurred forty years after Sunday services at the Lancaster Methodist Church with Fitzhugh Fulton (*"The Father of The Mother Planes"*) also in attendance. James Wood told his Shah story at a pool-side reception at a test pilot symposium in 1980. He had flown the Shaw of Iran out as the Boeing-707 command pilot as he escaped Tehran in January 1979.

Sparrow and Uncooked Toast

My early days at Edwards were full of excitement and learning. One famous "Black Day at Edwards" started around midnight and involved a runaway missile from the Navy China Lake Test Center (north of Edwards about 100 miles). The missile was the rumor mill then, but recent research says it was a WWII Hellcat configured for remote control, and it came out of Pt. Mugu headed for Los Angeles out of control. Air Force F-89s were called up from Oxnard AFB to intercept the Hellcat and chased it unsuccessfully over the San Gabriel Mountains, setting a small oil refinery on fire in Saugus and small forest fires with the rockets fired off the F-89s.

Fortunately, it turned back north, and the F-89s continued firing, pocking the pavement on Highway 138 with shrapnel, which landed through a living room wall in Palmdale (narrowly missing a lady sleeping on a couch). That morning, our carpool saw a towed Chevy behind a wrecker in front of the dealer. We wondered where was the source of the shrapnel embedded in the hood.

The day was just getting started, though, as about dawn, a Douglas F-4D test pilot had to eject north of the base. Our carpool would go to the commissary on the South Base, at that time, still in the old WWII wooden barracks buildings and offices, to have breakfast after the 45-minute commute from Lancaster. My toast was cooking when suddenly the power went out on the whole base: An F-86D model had crashed after cutting the power lines to the Air Force base at the foot of Lumen Ridge (the rocket base) in the rolling hills just east of Rogers Lakebed.

I was soon on my way to the crash site with an experienced GE rep that had worked on the F-86D in Korea. We heard the pilot had been trapped in the cockpit, unable to eject the canopy with the aircraft on fire. Luckily was rescued by another pilot, but then rumor had it that the Air Force Col. who had rescued him was court-martialed for some unknown offense. Chance Vought had a Regulus II missile in flight at the base power outage but had no auxiliary power up in time, and so lost their Reg II.

At the Convair plant in Fort Worth, TX, a year later, I was relating the sparrow story at dinner with the B-58 Air Force flight test crew pilot Fitzhugh Fulton, Lyle Schofield, and a third crew member. Because I was the local GE engine representative Lt. Col. Harry Trimble, as the B-58 Mach 2 Bomber Test Project Officer, asked me to go with him in the T-33 (T-Bird) two-place jet trainer to Carswell AFB in Ft. Worth. The flight test crew from Edwards was getting checked out in the B-58 at the Convair plant adjacent to Carswell.

Trimble required that I get certified in the altitude chamber before flying in the T-Bird, which required me to experience a simulation of an oxygen loss and the ability to put on a helmet and hook up the oxygen before going unconscious. We took the T-Bird to 40,000 ft. During the descent, I got a sinus block requiring treatment in the base infirmary (for which I received teasing) before we took the cross country. On the return trip later in the week, though, we encountered a thunderstorm at night and had to refuel at Davis Monthan AFB in Tucson. Harry got the teasing back that night after getting a sinus block during his descent and required treatment. He fooled me on the night's approach into Edwards; he broke through the heavy clouds over Boron north of Edwards, flying inverted so Boron's town lights were above me, disorienting me.

As I finished the story with the flight test crew at dinner, Trimble asked me if I wanted to hear the real story about the F-86D accident. We were all amazed as he told us about his direct involvement: He and a base photographer had taken off at dawn that morning on a photo mission in a T-37 jet trainer with side-by-side seating. Harry weighed over 200 lbs. It was a crowded cockpit with the photographer and camera equipment. On takeoff, Trimble noticed a sparrow flying around in the cockpit, so he captured it and put it in the zippered leg pocket of his flight suit, giving it a shot of oxygen from his mask hose throughout the flight.

During their high-altitude photography operations, the control tower asked Harry to make a fly-by over the accident. Over the burning wreckage of the F-86, Harry said he made a low pass and saw the pilot trapped in the cockpit, and he told the tower he thought he could land at the edge of the lakebed and run to the crash a hundred or so yards away. So he did and grabbed the camera from the photographer, who was tangled in his harness, and ran up the sand dune hill to the burning airplane. Harry hurled the camera through the canopy and pulled the pilot out.

He then related how the emergency crash crew arriving moments later had to pry his hands from the pilot's shoulders as they were locked during the trauma. When he returned to the locker room, he removed his flight suit and remembered the sparrow he had put into his pocket. He pulled the limp bird and laid it on the top of the locker. Upon returning from the shower, he found the sparrow had flown away (with quite a story to tell its offspring).

Lt. Col. Harry W. Trimble receives the Soldier's Medal from Brig. Gen. H. S. Holtoner, Commander, Air Force Flight Test Center, for heroic action at Rogers Dry Lake, when he dragged a pilot from burning aircraft after smashing the cockpit with movie camera[6]

On May 26, 1956, the Grumman Flight Test Operations at EAFB made the first flight of the F11F-1F *Super Tiger,* in which the J79 engine replaced the Wright J65 engine. I was assigned to support the Grumman team when I arrived at EAFB a few weeks later. The flights were usually scheduled at dawn to take advantage of cooler temperatures. For the altitude record attempts, a white-haired Navy Admiral ready for retirement would always be in attendance, even at 4 a.m.

On April 18, 1958, a test flight at Edwards AFB set a world altitude record of 76,938 feet. It wasn't long before the Air Force's F104-C surpassed the F11F-1F record by reaching 103,395.5 feet above mean sea level on December 14, 1959.

J79 afterburner development problems persisted, and two years later, as a flight test propulsion engineer, I worked for GE Flight Test, smuggling pilot burners and flame holders to Convair in Fort Worth, TX. They could not get the B-58s flying as not a single engine of the 20 installed in the five aircraft could get an afterburner light, but our GE Flight Test prototype parts worked well. I got on a step ladder in the pasture behind the flight line and monitored with binoculars, each light off with new parts. This accelerated progress as the Air Force contracts red tape kept the factory channels from delivering.

The local GE J79 service engineering representatives assigned to cover the Convair B58 program were respective to my undercover operations to get them flying without delay. They invited me to a poker game one night, knowing I was a greenhorn; as I had the luck of the draw, they were surprised when I cleaned them out.

The Chance Vought *Regulus II* missile (whose combat configuration included a nuclear warhead) was flight tested from EAFB. It was also tested at the Naval base at Point Mugu, next to Port Hueneme, just south of Oxnard, CA. The test flights would be local but extended up-range over California and Nevada into the target area at Dougway, UT. Operationally, it launched with a solid propellant booster rocket (of about 130,000 lbs. of thrust) attached to the bottom of the missile. It was flown from the submarine deck, providing the first nuclear strategic deterrence force for the United States Navy during the first years of the Cold War and especially during the Cuban Missile Crisis. Four Reg. IIs were hangared on the submarine deck.

Development testing started with the missile taking off from the lakebed runway, controlled remotely by a pilot using the remote control from the back seat of a T-Bird, with the T-Bird pilot flying in close formation alongside the missile. Later in development, the booster rocket was used both from a fixed launch pad and later off the submarine *Grayback* near Port Hueneme/Point Mugu. My experience on the *Regulus II* included searching for engine parts in the wreckage of several crashes in the bombing range where the missiles were lost on landing approach.

While working for GE at the south base in 1956 & 1957, I watched Ryan's X-13 VTOL fly several times. It was a small experimental jet fighter that could make vertical takeoffs and landings. The nose hook allowed it to hang vertically from a rod on a rotated flatbed truck that would position it to the vertical position for takeoff. Then the jet would complete its transition to the several-hundred-knot horizontal flight before transitioning back to vertical flight for landing with its nose hook on the vertical truck bed.

It was a clever arrangement with the pilot's seat rotating from a 45-degree position for vertical takeoffs and landings and then back to a conventional horizontal position for horizontal flight. The Ryan project engineer for the X-13 was Perry Row, a WWII B-17 pilot that flew in the Ploesti Oil Raids and would be my second boss at NASA in 1962.

Ryan's X-13 VTOL

Also, at that time, I watched the Air Force launch an F-100 jet fighter from a flatbed truck by using a large booster rocket (similar to that used by the *Regulus II* missile) attached to the bottom of the aircraft. Unfortunately, one of the few flights resulted in one of the release latches hanging up, causing the booster to hang down precariously from the bottom of the aircraft; the pilot ejected safely, but they lost the airplane.

Launch of an F-100 Jet Fighter From a Flatbed Truck

Soon after arriving at Edwards, I enrolled in USC Graduate Extension courses on the base. The first was on rocket propulsion, given at the rocket base on Leuhman Ridge, east of the main lakebed at Rogers Dry Lake. Depending on where you lived in Lancaster/Palmdale, the trip was about 85 miles. The professor was an old-time rocket pioneer who worked at the Army Rocket Arsenal Lab in Pasadena. We started with about 25 students meeting in the underground control room (still under construction) for Saturn's 1,500,000-lb.-thrust F-1 engine test stand. It was a great class, but as the commute was a killer, only three of us remained to get our credits.

On a Saturday, July 26, 1958, my boss grabbed me to join him in driving out to the sand hills between Rosamond and Rogers Dry Lakes. An F-104 had just crashed, and as we arrived, the helicopter was departing with the pilot's body, that of Cap. Iven Carl Kincheloe, Jr. As we surveyed the scene, the smoking wreckage revealed the zippers from his flight suit. That early version of the F-104 had a downward ejection seat, and Kincheloe had rolled the plane 90 degrees as he was about to crash and ejected out the side of the plane.

But his momentum took him directly into the fireball as the plane continued the roll to crash inverted with a full fuel load from takeoff moments before. He was destined to fly the X-15, which was to start flight tests by North American in 1959 (June 8, 1959). Unfortunately, the crash investigation revealed the cables attached to the heels of his flight boot to retract his legs against the ejection seat were mistakenly made way too long and resulted in his legs broken during the ejection; the fire was inescapable, so it was all over for him quickly.

Many fatal accidents occurred during advanced fighters' and aircraft's flight test programs. A memorable loss stands out with a tall Swedish Air Force test pilot in the early days of the F-104 program. He would run around the hangar with a midget on each shoulder, delighting all the maintenance crews and the crew of midgets who Lockheed had hired from Hollywood. The midgets would skinny up the narrow jet air inlets on the F-104 and inspect compressor blades for foreign object damage (FOD) between flights. It saved a lot of time, as engine removal was the only alternative to making the inspection. It was a sad day when he did not survive his F-104 crash.

Leaving GE Flight Test at Edwards
and Move to Point Mugu, August 1958

My young sister-in-law's family health (allergy) situation motivated us to relocate, leave the dusty desert climate, and live near the beach. We relieved her grandparents from providing her care as she entered high school. I had worked with Chance Vought at Edwards, supporting their J79 engine, then I accepted their offer to work at Pt, Mugu, the Naval Air Station just South of Oxnard, CA

I was out of town for a weekend a *Regulus II* incident caused a real challenge. The missile was aboard the submarine at the dock in Port Hueneme. The GE Rep aboard had authorized another start attempt on the J79 after a shutdown accompanied by an unusual clatter. Unfortunately, on the second try, compressor blades penetrated the compressor case, wrecking the engine and subjecting the missile, the submarine, and the dock to a major fuel explosion hazard.

As the propulsion engineer, I had to pursue my old employer (GE) with critical feedback on the ineptness of their technical representative on duty for his actions. The whole point was there should have been an inspection of the engine inlet that would have discovered the initial compressor blade failure, possibly from foreign object damage (FOD), before throwing more blades through the compressor case. Chance Vought's Flight Test Manager nicknamed me "Tiger" for handling the incident.

USS _Grayback_ Submarine

Returning to the dock, we learned the Navy would cancel the _Regulus_ II (along with the F8U-3) projects. I did not wait for the official layoffs I knew would be coming. Instead, I called my old family friend, Harold (Hal) Watson (Project Engineer for Talco Engineering Company in Mesa, AZ), who had an opening for a test engineer.

Talco merged with Rocket Power, a solid propellant rocket company that had moved from California, and thus my assignments changed, too. I was working on the Convair Supersonic Rotational B-seat, generally known as the B-Seat. I designed and built a water brake for testing the sled (powered by a new ROCAT ejection seat rocket on a track installed at Falcon Field, the airport site in Mesa). The water brake was relatively quick and cheap, consisting of Masonite panels of graduating heights placed as dams between the water-filled sled tracks. To give the desired deceleration characteristics to the sled, I installed a tractor seat upside down on the front to produce a rooster tail of water as it engaged more resistance going down the track. It worked!

Another project was developing the pyrotechnic gun that deployed the parachute and the field-testing the ideal powder charge for proper function. I was impatiently waiting for the theorists in engineering to calculate the appropriate charge of powder to load into the cartridge, so I began with small amounts, doubling it every test shot till I could get the slug attached to the parachute to exit out of the flap of the chute. Unfortunately, the last charge was too much, and I could hear the slug, which had severed the cord, whistling through the air while I ducked for dear life.

During the B-seat testing, many launched seats landed in the adjacent irrigated agriculture fields, which complicated recovery due to the muddy conditions. The maintenance chief, Corrigan (who had been a crew member on the B-29 that launched the unpowered Bell X-1 Rocket airplane before its delivery to NACA) had a wild idea: He proposed and got permission to install telephone poles supporting many cargo nets laced together in the field next to the Falcon Field airport site. The subsequent tries to hit the nets with the seat after a zero-zero ground firing (ejection test) never were successful; the seats always landed in the muddy field.

After a tornado hit the airfield, cleaning up and rebuilding the destroyed aircraft and all our outside cartridge/pyrotechnics (hazardous faculties) was a tremendous job. These were temporary structures on the airport ramps built of 2' x 4' lightweight framing covered by plastic, designed to immediately fail if any accidental explosion happened while loading the components used in the ejection seats. Fortunately, the tornado blowing all this apart did not result in explosions or fires, but it shut down production.

A new Talco assignment took me to work on the other side of the Salt River. Their newly- relocated manufacturing bunkers produced solid propellant sound rockets 10 or 12 inches in diameter and 10 feet long. Talco had hired college students from ASU as production workers, and my job was to supervise their work and ensure they followed safety procedures. Unfortunately, these unsafe habits were not working out well: One night during the recovery operation (solvent sloshing), about ten of the metal cases were rejected due to failed pours of the propellant (per X-Ray inspections), and because of continuing safety infractions, I determined to resign and leave my position. Unfortunately, my judgment was confirmed correctly, as there were fatal accidents within a few years.

1960's Decade: The Apollo Decade

In my first two years, NASA assigned me as the X15 flight operations propulsion engineer. In late 1962 NASA promoted me to the Lunar Landing Research Vehicle project engineer. 1963-1966. I left NASA in June 1966 and served as the Bell Aerosytems Lunar Landing Training Vehicle (LLTV) Technical Director till 1969.

My NASA Career

We still owned our home in Lancaster, so we moved back in April 1960. On May 1 of that same year, a few weeks after starting to work for NASA, Francis Gary Powers was shot down in his U-2 spy plane.

The airplane was never seen in public before the incident, developed by Lockheed Skunk Works in Palmdale, CA, as part of a classified USAF/CIA program. Arriving at work the morning of May 2, there were black limos in front of the main FRC building 4800, where we entered to report for work, and the center was a-buzz with activity. Per President Eisenhower's orders, a U-2 aircraft was brought from under wraps in the air force hangar, moved to the NASA hangar, and painted overnight with the NASA logo on the tail.

Joe Walker, our chief pilot, who had never seen the aircraft, had to taxi it for the press conference, which purported to demonstrate the aircraft was not a spy plane but was used for atmospheric research. The incident set back talks between Khrushchev and Eisenhower. Because Khrushchev had the pilot and aircraft with the cameras, per news reports, on August 17, 1960, Khrushchev pounded the table with his shoe, accusing Ike of lying about the U-2 getting lost over Russia while on a civilian mission.

The X-15

The X-15 was using two 6,000-lb. thrust XLR-11 rocket engines stacked vertically as interim rocket engines until the 60,000-lb. thrust XLR-99 was certified for flight by its delayed development program. The XLR-11 engines used ethyl alcohol and liquid oxygen (LOX) plus 90% hydrogen peroxide (H_2O2) for powering the turbo pump used to pump the fuel and oxidizer as the liquid propellants.

The XLR-99 used anhydrous ammonia (NH3) as fuel, liquid oxygen as the oxidizer, and 90% H2O2 to drive the turbo pump. The X-15 carried 1,000 gallons of LOX, topped off from the 300-gallon tank in the B-52 carrier aircraft as it climbed to about 40,000 feet for the airborne launch off the B-52's wing. This allowed it to cool another 30 or 40 degrees from the -297 Deg. F sea level temperature, packing more oxidizer for record performance for flight records. The fuel tank carried 1,400 gallons of NH3 for the XLR-99 and lesser amounts of ethyl alcohol for the XLR-11 engines.

<u>Interim XLR-11 rocket engines installed in the X-15</u>

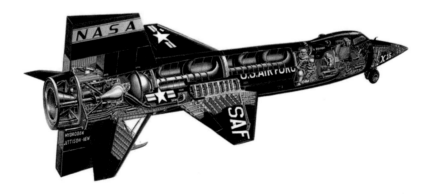

<u>Cutaway view of X-15 with 60,000-lb. thrust XLR-99 engine</u>

The XLR-11s made the first 25 flights (6/8/59–11/15/60 on both X-15 #'s 1 and 2. After that, the first XLR-99 engine flew on X-15 #2. One of my first assignments as the X-15 propulsion engineer in Flight Ops was to accompany an inspector on the rocket test stand to inspect a hot firing of one of the four chambers on one of the XLR-11 engines. It was running at 1,500 lbs. full thrust about a foot away from my face; we were looking to see if the LOX check valve weld was leaking. The LOX was at about -310 deg. F, but only safety glasses and ear protectors were worn, a far cry from today's more exhaustive safety practices.

The XLR-99, a cutting-edge and complex rocket engine, was the first of its kind designed to be shut down and re-started in flight and be throttleable from 50% to 100% thrust. My job was to develop supplemental training aids to be used by our operations engineers, technicians, and inspectors. Our Flight Operation Engineering Branch had a draftsman named Harry White who had been there through the X-1 days. He had emigrated from South Africa with a significant accent and a positive work ethic, commuting in his meticulously-kept Porsche sports car weekly from his home in Apple Valley.

Harry prepared multiple large-scale drawings (about 4' wide x 3' high) of the engine's control systems diagrams for each mode of operation, color-coding all fluid and electrical systems to depict the engine operation. These drawings were mounted on hinged plywood panels and set in a large box on casters used in classrooms and aircraft hangars. One of the inspectors, John Reeves, went further and developed an analog checkout box to assist in checking out the engine control system in the aircraft as part of the pre-flight maintenance inspections, an excellent example of the large expenditure of considerable man-hours to produce complex multi-colored training materials in 1960 compared to 21st-century technologies

On the 2nd ground run of the XLR-99 engine on X-15 #3 (June 8, 1960), the ammonia tank pressure regulator and the relief valve failed. The tank exploded, shooting the 3,600-psi helium tank from the core of the forward LOX tank into the H_2O_2 tank, resulting in another terrific explosion that destroyed the airplane. The fuselage severed at the aft edge of the wing. The pilot in the cockpit, Scott Crossfield, shot forward 20 or 30 feet while the tail assembly was blown into the lakebed by the explosion.

X-15 #3 after the fuel tank explosion severed the aircraft

Returning to Edwards from the XLR-99 30-day training class in the Danville, NJ., Reaction Motors plant, Jim Love assigned me to closely follow all the investigation activities and development of North American Aviation's corrective actions and report to him daily. This challenging experience showed me the importance of due diligence and a great deal about the importance of objective and independent reporting on complex technical issues. The chronic failure of the responsible components was a sensitive and embarrassing issue for the company. The accident drove the imperative efforts to fix the design and quality, even though another problem caused the explosion. The high-back pressure on the relief valve outlet port tubing had been inserted into a water tank to keep the vented ammonia from escaping around the test stand, the usual procedure. The chronic relive valve failures, though, this time caused the tank to rupture. Later, Jim Love sent me to a one-week rocket propulsion seminar at UCLA. The conference was an excellent experience for learning what was happening with rocketry in those early days after Sputnik.

Tackling Bureaucracy
(When you detect unacceptable contractor support/bureaucratic inertia, make your case with facts professionally)

Ground testing and flight operations were fast-paced, with X-15 #3 rebuilt after the test stand explosion and the other two X-15s using up our inventory of eight flight engines. In addition, the XLR-99 had countless component failures requiring replacements and re-testing both in the Air Force maintenance shop and re-firing the engines in the test stand. Many of these failures were chronic and needed long lead times for Reaction Motors and the Air Force to force vendors to provide fixes.

My job was to complain and complain and dig for answers to dispute the "official" disposition coming from the bureaucrats. I was a thorn in the side of field reps for Reaction Motors (they nicknamed me "*Murphy*" in frustration), but I got things moving as I learned how to get accurate facts and move my feedback to higher management levels. An example was the corrosion of the pressure switch contacts in the control box, requiring a design fix of replacing the contacts with gold to eliminate failures. In addition, I had to contact the vendor directly to discover the regular false status reports for missed deliveries of the new parts represented by Reaction Motors and the Air Force. These engine component failures were causing frequent and severe delays in the availability of flight engines for aircraft operations and flights. However, this experience in breaking "log jams" usually made my future challenges manageable.

In 1962, frequent explosions during flights occurred in the instrumentation tubing, transmitting the XLR- 99 engine two-stage igniter pressures to the pressure transducers after engine shutdown. Reaction Motors and the Air Force determined the cause was a bad batch of "*LOX Safe*," a lubricant used in the rocket engine plumbing/piping assembly. Subsequently, all engines were torn down and rebuilt with new lubricant. The fiasco set the flight program back several months, but the problem persisted. I was convinced we should not fly till the issue got resolved, but Paul Bikle, our center director, disagreed. I convinced him at least we should construct a steel shield and install it in the area where the lines were exploding in the heart of the engine to reduce the risk of catastrophic engine damage. Our NASA machinists, as usual, came up with an innovative design, machining pieces of thick curved steel small enough to insert into the cramped area around the exploding tubing and assembling them with clamps, forming an almost spherical steel ball as a shield. We flew with these till the Air Force had an engine explode in the test stand, forcing an accurate diagnosis of the real problem.

X-15 Movie (1961)

Frank Sinatra's film production company received permission from NASA to produce a feature film using the X-15 as the subject for a dramatic story featuring Charles Bronson and Mary Tyler Moore in her first leading role. NASA allowed them one week of filming on-site at Edwards, with limited access to the facilities and personnel. Milt Thompson was soon to fly the X-15 and was assigned as the technical advisor to assist in the production. The film executives were invited early to witness an X-15 rocket ground firing of the XLR-99 engine, tied down in the test stand area. Milt had them with him in NASA 9 Mobile. A communications van for flight operations was parked about 50 yards outside the test area for the firing.

NASA Mobile Control Van

I was in the closest pillbox to them, left and aft of the aircraft, about 20 feet to the side of the rocket exhaust. The underground concrete pillbox had about a foot thick of multilayered explosion-proof glass above the ground level for viewing the aircraft and rocket plume. It had a heavy blast-proof lid to access the ladder accommodating five men: an aircraft mechanic, the crew chief, myself, the operation engineer, and two inspectors, fully suited with oxygen backpacks. As Milt explained the rocket run to the movie execs, the engine controls detected a problem and shut the engine down. The two inspectors exited our pillbox and climbed onto the aircraft to find any abnormal conditions.

The crew chief, Larry Barnett (about 6' 4"), had climbed to the top of the pillbox ladder to watch the crew on the aircraft, and I stayed at the bottom of the pillbox with my headset plugged into the communication net. The communications net included the three pillboxes (one at the nose of the aircraft and the third opposite the one I was in) and the pilot in the cockpit. Unfortunately, after a few minutes of the inspection, a massive ammonia leak occurred in the aircraft tank vent's ground support system. The cloud of 90% NH3 blew at our pillbox, knocking Larry off the ladder and falling on me as I pulled my filter mask over my headset (only suitable for up to 3% NH3).

This level of ammonia exposure can be fatal, shutting down the respiratory system immediately. It seemed that Larry and I passed each other several times, scrambling up the six-foot ladder, and when we reached the top, we jumped off the top of the pillbox as far away as we could, leaping every other step to escape the heavy gas. As we were headed towards the van with Milt and the film executives, I came to the end of the communications cord on the headset, and it sheared off, shutting down the entire communications system. I never missed a stride, though, doing a complete flip in the air from the force of the cord connector shearing off. Stan Novak grabbed his filter mask and ran into our pillbox. Then, restoring the intercom, he ran back out of the cloud. His admirable actions restored communication for the other ten people. Meanwhile, Milt had some explaining to do for the film executives. Thirty-two years later (and about a month before his roast), Milt autographed his new book *At The Edge of Space* for me, having me read pages 79 and 80, which detailed this event.

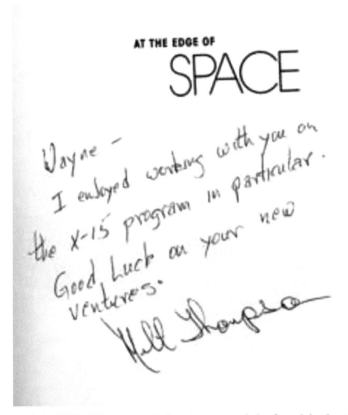

Autograph in Milt Thompson's book a month before his death, 1993

Six plume signatures of increasing Mach numbers

"It was as humorous as any Charlie Chaplin or Marx Brothers routine I had ever seen," he wrote. Before those two sentences, though, he had fictionalized the description of the incident. He claimed we had opened the lid of the pillbox during the rocket firing and had run through exhaust flame. The acoustics alone would have been fatal to all of us, let alone incinerated in the exhaust flame. Plus, we were running away on the left side of the plane. We could not have run through the rocket plume as he claimed we had opened the lid of the pillbox during the rocket firing and had run through exhaust flame.

A few weeks later, I was chosen among several operations personnel to appear in the movie Mission Control. In the photo. I am sitting next to my NASA Colleague Bill Albrecht. Before the scene's shooting, the director requested we do whatever we would do in a real mission. I asked for a Dutch Master cigar which they brought me from our cafeteria. Halfway through the shooting, a lunch break in the hangar was taken with prime rib. Before the shooting re-started, the continuity lady said, "Ottinger, cut that cigar in half to match the before-lunch scene." Hence, I just sucked the cigar in for the same effect and heard no more from her. Later, Sondra and I got to go to the premier showing in Hollywood. I was off duty when they handed out the dollar bill to each of us for participating, so they still owe it to me.

<u>Mary Tyler Moore and me with my cigar.</u>

1962 NASA FRC Gets a Visit From the Pregnant Guppy

This unique modified cargo aircraft was quite a site as it parked near our building near our hangar. Little did I know one of these would be transporting my #2 LLTV from Ellington Field north of the NASA MSFC in Houston to the NASA Langley Wind Tunnel in Virginia in January of 1970. About the same time in 1962, a demonstration of a pilot rescue system with a low-flying C-130 snatching a man under the cable strung between two poles, showing an unlit cigar before the snatch and the cigar lit as he arrived in the C-130 cargo bay after the snatch. I'm not aware that the rescue system was ever used in combat.

LLTV #3 Loading Into Pregnant Guppy for the trip from Ellington Field to Langley Wind Tunnel

Clay Lacy and his Apollo Salvation

https://www.claylacy.com/wp-content/uploads/2016/09/Lucky-Me-Book-about-Clay-Lacy.pdf

The Pregnant Guppy was the first of a unique aircraft family that has been transporting outsized cargo for almost 40 years. Their loads were cargo that couldn't be transported any other way or only with great time and trouble. The Pregnant Guppy had a humble beginning on the proverbial cocktail napkin. One evening friends Jack Conroy, Lee Mansdorf, and others were discussing the problems NASA was having transporting the rocket booster stages aboard ships through the Gulf of Mexico. Mansdorf had recently purchased several surplus Boeing Stratocruisers but wasn't sure what to do with them. Conroy figured they could take one of the Stratocruisers, enlarge the fuselage big enough to hold a rocket booster, and contract with NASA to fly the boosters from California to Florida. The napkins were flying that night! Some of the ideas discussed included opening the top of the fuselage hinge. The cargo would then be lowered in with a crane. This idea was scrapped early on for obvious reasons.

From Aero Spacelines Super Guppy Appendix A Page Page 385

http://www.allaboutguppys.com/pg/377pgf.htm

In late 1962 the Pregnant Guppy flew to NASA's Flight Test Center at Edwards AFB, California, for more intensive flight testing. This testing continued into early 1963, with the only mishap being during the water ballast testing the water system malfunctioned and flooded the floor of the airplane. The pilot quickly landed with a very damp and irritated test engineer in the Guppy's lower deck.

By this time NASA management began pushing for the use of the Guppy as soon as possible with numerous letters and phone calls going to FAA officials. NASA was planning the first two-stage launch of the Saturn I vehicle, the SA-5. Because of time lost due to testing problems with the SA-5's first stage, time was becoming critical, and the Guppy promised to fly the S-IV-5 stage to the Cape in eighteen hours as opposed to 18-25 days if shipped through the Panama Canal and the Gulf of Mexico.

Beatty Nevada Trips

<u>C-130 used for X-15 High Range Mission Support</u>

Usually beginning around 3 AM at base operations, eight or ten NASA and Air Force mission control engineers would play card games in the barbershop, waiting for the AF C-130 to arrive at the loading ramp to pick us up. The communications technology and radar systems required mission control to be reasonably close to the launch sites: NASA One at Edwards, Beatty mid-range for Mud Lake launches, and Ely, NV, close to the north end (Wendover, UT) of the 400-mile-long by 50-mile-wide High Range setup for X-15 flights.

I never went to Ely, but many times was in the Beatty High Range Radar and Telemetry facility serving as the propulsion member of mission control. We covered the flights launching from Mud Lake till they progressed south within range for NASA One to take over control at Edwards. The C-130 would always land on a 2,000-ft. dirt strip located not too far from the town of Beatty, but we usually needed ground transportation, two or more carryalls. So we left the dirt strip, traveled over a winding dirt road over a mountain into Beatty, went north for 20 miles on Highway 95, and then two miles on the gravel road to the top of the ridge west of the highway to the highway station.

On the C-130 we boarded at Edwards was a fire truck, and a sometimes-over fueled jeep that would slop gasoline around on the aircraft's floor during turbulence. As no seats were available, the passengers, about 15 or 20, usually stood holding onto straps hooked onto the fuselage. This group included the mission control team, paramedics, flight surgeon, and technicians for the jeep, equipped with X-15 rocket system safing equipment. Once, an alert passenger knocked a cigarette out of a newbie's hand standing in the gasoline puddle before he could light it. Some F-104 weather reconnaissance pilots would find us on the mountainous dirt road just after leaving the C-130, approach behind our vehicle flying flow, rotate right over our vehicle into a vertical climb, and light their afterburner for a "welcome to Beatty!" gesture. As we passed the local brothel just north of Beatty on Highway 95, we all checked the number of Greyhound buses parked there, teasing the local NASA technician stationed at the site about the brothel.

Bruce Peterson, a NASA test pilot assigned as an X-15 chase and Lifting Body test pilot (also a board member of the nonprofit PAT Projects, Inc. I founded in 1994) in the 1990s, told me a story. On an F-104 flight, maintenance crews were working on the X-15 emergency lakebeds' runway markings, and he buzzed them with a low supersonic flyby. The crew got revenge by taping the radio mic switch down, so on the next pass, Bruce got the sonic boom back in his headset, ending his antics. Bruce was the pilot that crashed end over end in the opening scenes of the TV series *"The Six Million Dollar Man."* He lost his helmet in the rollover, and Bruce suffered severe facial injuries and eventually lost an eye due to a staph infection. He never would allow plastic surgery, and so wore a large black eye patch for the rest of his life.

Flight cancellations for weather or aircraft problems often meant rescheduling a flight for the next day. So our mission control team would fill up nine-passenger black & white NASA station wagons and head about 130 miles south to Las Vegas, an attractive alternative to boring Beatty. Only a couple of rooms were rented at the casinos. So we all shared rooms and spent much time in the lounge shows. Free lounge shows were plentiful in those days. Winners at the tables would pick up the tab at dinners, and we would all head back early the following day to meet the new flight schedule, some of us more alert than others, though this never comprised our performance, as it turned out.

On one trip, my boss, Perry White Row, drove at least 80 mph. I was with Dick Day in the third seat facing to the rear with the tailgate down, our feet stretched out on the tailgate, pitching spent beer bottles out to our left (not to be proud of today) when an explosion caused Perry to head for the dirt with a maximum braking maneuver. He thought he had a blowout; it turned out that one of the beer bottles had hit a road sign. Dick was our X-15 stability and controls specialist and flight planner and shortly after joined our former Center Director Walt Williams at the Manned Spacecraft Center (now the Johnson Spacecraft Center). Soon after Dick arrived, he noticed Neil Armstrong's rejected application for the second round of astronauts (it had reached a few days late). Having worked with Neil for many years at FRC, he moved the application into the accepted box. I'm sure he did not quite realize how he was impacting history.

Ham the Chimp

On February 1, 1961, the day after the Mercury flight of Ham the Chimp required by the medical team before Alan Shepard could make the first flight on Mercury-Redstone), during an X-15 post-flight briefing, Walt Williams told us about Ham's flight.

Ham before the Mercury flight on the Redstone Rocket, Jan. 1961

He told us they had trained Ham and the backup chimps to use a keyboard sequence they would punch to receive banana chips. Ham eventually learned he could take a shortcut from how he trained and still get his banana chips. They also exposed the chimps as much as possible to the flight hazards to alleviate their fear. One big surprise occurred when they observed on film that he got scared when the escape tower left the capsule with a blast after the booster successfully dropped off, a hazard they had omitted in training. Ham immediately went back to the former method of getting his banana chips. Unfortunately, the official NASA reports contained no mention of this story.

Ham's flight turned out to be longer than planned by 130 miles, and by the time they recovered him in the capsule, the water had climbed to the edge of the couch he was strapped to; the rescue was just in time. He survived the mission and lived in two different zoos till he was 26 years old. Walt later commented that the doctors were worried about humans' ability to survive in space and considered a hundred chimps for testing. But Walt got Joe Walker (who already had many X-15 flights experiencing high heart rates at launches and landings) to convince the doctors that these measures were unnecessary since he never fainted or passed out. So Ham's flight before Alan Shepard and Enos's flight before John Glen were the only two chimp flights made in the Mercury program.

The X-15 Research Airplane Program at Risk

We were running out of thrust chambers for the engines as they lost their ceramic (Rokide) coating on the 3/8" stainless steel cooling tubes (welded together to form the rocket chamber). The Rokide (zirconia or zirconium oxide) insulated the chamber from the high-velocity combustion flame of 5,000 deg. F at 600 psi.. The 8.5-inch diameter chamber throat suffered the bulk of the ceramic coating loss at the early stages of operation, eventually causing failure or cracking of the stainless steel tubing. In addition, the raw 90% NH3 fuel (at -49 Deg. F) used to cool the chamber would leak into the combustion process and, if left unchecked, could destroy the chamber. Reaction Motors applied the Rokide at their factory in New Jersey before the injector was welded onto the chamber assembly. The chamber repair required the engines to be returned before reuse.

The engines were completely torn down, the injector cut off from the chamber, and the rod-fed flame-spray gun sprayed on each end of the open chamber to strip the old coating and re-apply the new coating with inferior life expectancy. The injector face was then re-welded to the chamber. The entire process took almost a year and approached a million-dollar repair bill. Moreover, with the terrible early failure rate in the field and only eight flight engines supporting three X-15s, the whole flight program was in grave danger of progress being painfully slow. Our NASA FRC Director, Paul Bikle, assigned Al Covington from the research division under Don Bellman and me in the flight operations division to focus on a more satisfactory solution. Al was a member of the base flying club and wanted to rent a Navion for our visits to various process and material research facilities. However, when we attempted this, the DC generator on the Navion failed from Edwards to San Diego in bad weather. As a result, we had lost all communication and instruments, and having landed in San Marcos, we re-charged the battery and returned to Edwards on battery power alone.

After that, using a government car and commercial airlines, we visited research labs at Rocketdyne in Simi Valley, Aerojet in Sacramento, and Solar Turbines in San Diego. We discovered their work on developing gradated ceramic/metallic coatings using plasma arc deposition guns (about 20,000 deg. F), which were much smaller than the flame spray rod-fed guns used by Reaction Motors. I immediately recognized the potential to adapt this plasma arc deposition system for a rapid field repair of the many premature chamber coating failures. So I began working with our machine and sheet-metal shop personnel at NASA FRC.

Bellman and Bikle gave us the green light with maximum center priority to proceed. Working with the drawings for the plasma-arc gun and the rocket chamber, I led a team of nine shop people to repair a fully-assembled engine and complete a stripping and new coating operation in one week (the only way to keep them flying). We had to keep the gun perpendicular to the contour of the chamber, precisely controlling the distance from the tubes and moving across the surface through the length of the chamber at research lab-dictated surface speeds. We all knew we needed to make a pantograph machine within the first few days, and we were off and running. Everyone started with their expertise in designing the machine and building the components. First, I had to get the Air Force Col. in charge of maintenance to loan us a jet engine rollover stand. I stopped a Little League game to get him to sign off the loan papers from his position behind the pitcher.

We had just enough room to mount the rocket engine, fully wrapped in plastic on one half and the pantograph mechanism on the other, onto the rollover stand. Two-inch-thick hardwood planks were glued together and turned on a big lathe to form the plug duplicating the chamber geometry. The plug enabled the accurate fabrication of the metal hardware. The machine shop made the mechanical linkage to hold the plasma-arc gun positioned by the cam linkages assembly aft of the engine to follow the chamber contour defined by the hardwood plug. 12-hour shifts, seven days a week, for about seven weeks for all 10 of us brought it ready to travel to our first trial repair.

The Air Force program for gradated coating research worked with Plasma Kinetics and later the Air Force Power Plant Branch at Edwards to supervise the coating formulae. Still, NASA (my team) supplied the field repair pantograph machine in a couple of months, the time it would take the Air Force to get started on a new project. We had a NASA photographer come with us to document the first coating application. We were there probably a week or so, successfully demonstrating to the Air Force that they had a field repair capability ready to fine-tune to keep the X-15 flight research program going.

As they were constantly needling me, my two agitated Reaction Motors field reps named the machine "Murphy's Machine." Many years later, however, one of the Reps, Billy Arnold, acknowledged it was a successful solution to their critical engine reliability problem. NASA applied for a patent, but it was rejected due to "prior art," based on a 1929 patent to coat light bulbs with the frosted coating pantograph machine. The NASA Awards System decided this effort deserved a $10,000 award, a far cry from the traditional formula based on future innovation savings. I told my nine shop people we should just each take the $1,000 and forget it.

The 60,000-lb. XLR-99 thrust chamber Rokide Erosion

Plasma Kinetics, an Air Force Vendor in Los Angeles

A One-week repair (replacing the one-year factory repair)

My Apollo Choice, A Career Milestone

A year later, when it came time to assign new projects in my operations engineering branch, I was offered either the operations project engineer Dyna-Soar program or the LLRV.

<u>Artist drawing of the X-20 Dyna-Soar</u>

Fortunately, I chose the Apollo LLRV. Ironically when I was at Bell AeroSystems in 1963, using General Walter Dornberger's (von Braun's boss in Germany) office for my private calls to NASA, I saw a painting of his X-20 on the wall. Dornberger conceived the Dyna-Soar "Skip Glider" spaceship, later known as the X-20, contracted to Boeing. Secretary of Defense Robert M. McNamara (referred to as *"Mac the Knife,"* the name of a popular song) canceled the project in 1963. The painting on his office wall was a beautiful artist drawing with a plaque saying, *"Dyna-Soar, born 1952, Walter Dornberger, died, Mac the Knife 1963."* While viewing the painting, I had an uneasy premonition about the big-stakes world of aerospace R&D financing. I was a bit concerned about the total development costs required for the LLRV. The Dyna-Soar painting reminded me that FRC was playing in an ocean with some very big fish

Richard E. Day 1917 - 2004

On one trip from Beatty, Nevada, to Las Vegas, my boss, Perry White Row, drove at least 80 mph. I was with Dick Day in the third seat facing the rear with the tailgate down. Our feet stretched out on the tailgate (embarrassing today), pitching spent beer bottles out to our left when an explosion caused Perry to head for the dirt with a maximum braking maneuver. He thought he had a blowout; it turned out that one of the beer bottles had hit a road sign.

Dick was our X-15 stability and controls specialist and flight planner. Shortly after joined our former Center Director, Walt Williams, at the Manned Spacecraft Center (now the Johnson Spacecraft Center) in 1964. Soon after Dick arrived, he noticed Neil Armstrong's rejected application for the second round of astronauts (it had reached a few days late). Having worked with Neil for many years at FRC, he moved the application into the accepted box. I'm sure not quite realizing how he was impacting history.

It is appropriate to talk about Dick's abundant contributions to early supersonic aerodynamic solutions to early fatal X-Plane accidents.

Coupling Dynamics in Aircraft: A Historical Perspective

Richard E. Day
Dryden Flight
Research Center
Edwards, California

Day's NASA Report Documenting His Solutions for Early Fatal Supersonic Crashes

Dick Day left NASA in April 1964 to join Walt Williams at the Aerospace Corporation working on many new Air Force space projects, including the Manned Orbital Laboratory, until 1975, when he returned to working for the NADSA DFRC. I visited Dick at his home in Palmdale in the late 1990s. I helped him adapt to the then-current spreadsheets and word processing as he documented his FAA certification work for the owners of the Pregnant Guppy. Unfortunately, his health started declining when I was with him, though he fought through it till 2004. The following *NASA DFRC NASA Express* mentions his pilot's license was issued in Canada in 1938.

C. Wayne Ottinger

Day leaves behind a rich legacy

By Sarah Merlin
X-Press Assistant Editor

At his death on July 4, NASA engineer and researcher Richard E. "Dick" Day left behind a legacy that touched virtually every manned space program in U.S. history.

Day, who was 87, had been a member of the NASA family since leaving college in his native Indiana after World War II. From a long list of achievements – which included pilot, flight instructor and flight test engineer along with significant roles in multiple high-profile research projects – he is perhaps best known for his pioneering work with flight simulation.

Retired Dryden engineer and test pilot Bill Dana enjoyed a personal and professional relationship with Day that spanned more than four decades.

"We opened a relationship when I first came here (in 1958) – he was my very first boss at NASA – and it lasted until the very end," recalled Dana, who as a young pilot was among those who worked with Day and his early flight simulator.

"Dick was the warmest person. And he combined that warm personal friendliness with great technical capabilities. I will miss him a lot."

Day began his aeronautical career in the pre-war Civilian Pilot Training program, receiving his pilot's license in 1938. In 1940, he volunteered to join the Royal Canadian Air Force and graduated from the RCAF aviation cadet program, in which he also served as a flight instructor after graduation.

Following the 1941 attack on Pearl Harbor, Day was allowed to transfer from the RCAF into the U.S. Army Air Forces and flew combat missions over Europe in B-17 and B-24 bombers. After the war, he earned a bachelor's degree in physics and mathematics at Indiana University in 1951 and went immediately to work for NASA's predecessor, the National Advisory Committee for Aeronautics.

Day's work at what began as the NACA High Speed Flight Research Station on Edwards Air Force Base and eventually became, evolving through various incarnations, Dryden Flight Research Center had an auspicious beginning with early assignments on the X-1 rocket research aircraft and the XF-92 delta-wing projects. Then in 1953, Day found himself with access to the U.S. Air Force's analog computer at Edwards. He programmed the computer with the characteristics of an airplane – as those characteristics had been recorded in flight – then added a control stick and a cathode ray tube, which served as the pilot's display. With this configuration, he effectively became the first to construct a rudimentary flight simulator.

Day's handiwork became an integral part of research in the X-2 program. After a 1956 crash in which both pilot Milburn G. "Mel" Apt and the X-2 aircraft were lost to an uncontrolled tumble, Day was able to recreate with the simulator the conditions leading to the crash, defining new controllability prediction techniques that would prove invaluable in preventing future accidents.

"Mel crashed because they hadn't figured out yet what caused those kinds of (flight) conditions to occur," said Dana. With his simulation research, "Dick gave them the ability to identify those conditions so they could prevent crashes like that one and others from ever happening again."

Concurrently with his work on the X-2 program, Day became involved in similar testing with the Douglas

E3733   NASA Photo

In a career that spanned more than four decades, former Dryden engineer Richard Day was a pioneer in both flight simulation and in the aircraft design concept of roll, or inertial, coupling. The Indiana native, shown here instructing colleagues in use of the inertial coupling demonstrator he built, came to work in 1951 for NASA's predecessor at Edwards Air Force Base, the National Advisory Committee for Aeronautics. During his tenure at what eventually became the Dryden Flight Research Center, his work influenced virtually every U.S. manned spaceflight program. Day died July 4.

X-3 jet-powered aircraft, designed to accommodate research at speeds around Mach 2. Simultaneous research with Air Force F-100As offered still another environment for Day's simulator work with roll coupling and resulted in a flight configuration that was to give the F-100 aircraft series satisfactory resistance to roll coupling divergence.

But his invention and his capabilities were never put to better use than in the X-15 program, in which Day planned the legendary aircraft's envelope expansion with both the interim and production engines. Using the simulator, he analyzed the X-15's stability and defined key aspects of the aircraft's performance at high speeds.

His thorough understanding of the X-15's capabilities put him in a unique position to serve with mission planning, pilot training, development of operational procedures and energy management studies. He also possessed the right combination of engineering and communication skills – the latter hearkening back to earlier work as a flight instructor – to allow him to work closely with the X-15 pilots as they learned to fly the ground-breaking aircraft.

In 1962, Day had yet another occasion to train young pilots when he left California for the Manned Spacecraft Center in Houston (now Johnson Space Center) to become assistant division chief for astronaut training on Project Mercury. There, in addition to training astronauts on the simulator, he developed training programs, wrote skills tests for the astronauts and served on the astronaut selection board.

From 1964 to 1969, Day worked for the Aerospace Corp. on the Manned Orbiting Laboratory project, again as an astronaut trainer. But in 1975, he was asked to

See **Day**, page 12

EC03 0110-1   NASA Photo by Tom Tschida

E1841   NASA Photo

Above, Day discusses use of the inertial coupling demonstrator in a 2003 interview at his home. In 1953, he was the first to develop a rudimentary flight simulator, at which he is pictured here. Day's work helped engineers develop controllability prediction techniques that were invaluable in preventing aircraft accidents.

Teamwork Abounds

During an X-15 flight attempt in August 1961 in hot temperatures, I was in the NASA One control room when the telemetering signal was lost during the B-52 climb out to launch altitude. The responsible engineer convinced Joe Vensel, our Flight Operations Director, that a connector had likely come loose in the equipment bay behind the cockpit. If Joe approved, they could land, he could correct the problem, and they take off again and resume the mission. Joe agreed, and the B-52 came into land. The landing was heavy, with a full load of B-52 fuel and X-15 propellants. On a hot day, it was too much for the tires on the B-52 to take. By the time it rolled to a stop and turned off the end of the runway, every tire had blown, making the X-15's lower ventral at the tail end almost touch the tarmac. Smoke was coming out of the equipment bay behind the cockpit. Boy, did it light a fire under all of us in mission control. We all knew we had a potential disaster as the Air Force fire trucks and emergency crews extracted the pilot right away, with a NASA crew chief replacing the pilot to do whatever he could to manage what was needed. We had to unload all the X-15 propellants, including 300 gallons of LOX in the B-52 used for top-off at altitude.

I immediately went to my office and got help loading my four-drawer file of drawings into a pickup and getting down to the aircraft. The runway location was several hundred yards away from the servicing area where the propellant loading had been done. So all the support from there had to be quickly adapted to the site adjacent to the end of the runway. It was quickly determined that we could not open the normal access panels to de-fuel the X-15 as they were stress panels unable to be removed with a full fuel load over 15,000 lbs. Every step was carefully planned, reviewed, and critiqued by the whole team of engineers, mechanics, and inspectors. The entire process continued all night with the portable lighting and area access control commensurate with the obvious hazard. The fact that we got the job done without any explosion, fire, or injuries was a real testament to the teamwork and skill of the entire crew. The electrical fire in the equipment bay turned out to be less scary than it could've been, but it taught us all some great lessons. For the rest of the X-15 program, no B-52 landings were ever made without jettisoning all propellants before landing.

From the X-15 to the Apollo Program

In the spring of 1962, I accepted the assignment as the project engineer on the LLRV and began working with the project office managed by Don Bellman and Gene Matranga. They had been working with Bell Aerosystems, who had proposed a gimbaled jet concept to address the piloting challenges of lunar landings (1/6 lunar gravity and vacuum). My choice for this assignment was to be the most important decision ever made in my professional career, as it impacted my entire life from 1962 through 1968 and from 2007 to the present. This was one of the major decisions of my life.

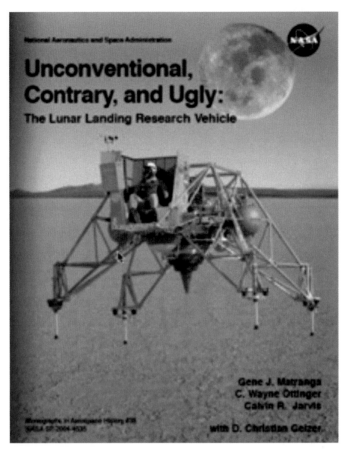

NASA Monograph I co-authored and produced

Foreword by Neil Armstrong

"Landing a craft on the moon was, in a number of ways, quite different from landing on Earth. The lunar gravitational field is much weaker than Earth's. There were no runways, lights, radio beacons, or navigational aids of any kind. The moon had no atmosphere. Airplane wings or helicopter rotors would not support the craft. The type of controls used conventionally on Earth-based aircraft could not be used. The lack of an atmosphere also meant that conventional flying instrumentation reflecting airspeed and altitude, and rate of climb and descent, would be useless because it relied on static and dynamic air pressure to measure changes, something lacking on the moon's surface. Lift could be provided by a rocket engine, and small rocket engines could be arranged to control the attitude of the craft. But what trajectories should be selected? What type of steering, speed, and rate-of-descent controls should be provided? What kind of sensors could be used? What kind of instruments would provide helpful information to the pilot? Should the landing be performed horizontally on wheels or skids, or vertically? How accurately would the craft need to be positioned for landing?

What visibility would the pilot need, and how could it be provided? Some flight-test engineers at NASA's Flight Research Center were convinced that the best way to gain insight regarding these unknowns would be the use of a free-flying test vehicle. Aircraft designers at the Bell Aircraft(Aerosystems) Company believed they could build a craft that would duplicate lunar flying conditions. The two groups collaborated to build the machine. It was unlike any flying machine ever built before or since. The Lunar Landing Research Vehicle (LLRV) was unconventional, sometimes contrary, and always ugly. Man y who have seen video clips of the LLRV in flight believe it was designed and built to permit astronauts to practice landing the Apollo Lunar Module (LM). Actually, the LLRV project was begun before NASA had selected the strategy that would use the Lunar Module!"

Our LLRV concept at NASA FRC and Bell was ahead of its time, even ahead of the Apollo concepts of how the landings were to be made, but we all knew that flight research and pilot training would be critical to the success of the landing mission. We just lucked out that the LOR mode finally chosen matched our LLRV/LLTV concepts.

John Cornelius Humbolt
<u>(April 10, 1919 - April 15, 2014)</u>

was an aerospace engineer credited with leading the team behind the lunar orbit rendezvous (LOR) mission mode, a concept that was used to successfully land humans on the Moon and return them to Earth. This flight path was first endorsed by Wernher von Braun in June 1961 and was chosen for the Apollo program in early 1962. The critical decision to use LOR was viewed as vital to ensuring that Man reached the Moon by the end of the decade, as proposed by President John F. Kennedy. In the process, LOR saved time and billions of dollars by efficiently using existing rocket technology.

The Lunar-Orbit Rendezvous Concept

"Whether NASA's choice of LOR would have been made in the summer of 1962 or at any later time without the research information, the commitment, and the crusading zeal of Humbolt remains a matter for historical conjecture. His basic contribution, however, and that of his Langley associates who, in their more quiet ways, also developed and advocated LOR, seem now to be beyond debate. They were the first in NASA to recognize the fundamental advantages of the LOR concept, and for a critical period in the early 1960s, they were also the only ones inside the agency to foster and fight for it. The story of the genesis of LOR underscores the vital role occasionally played by the unpopular opinion. It testifies to the essential importance of the single individual contribution even within the context of a large organization based on teamwork. And it demonstrates the importance of passionate persistence in the face of strong opposition and the pressure for conformity.

Thousands of factors contributed to the ultimate success of the Apollo lunar landing missions, but no single factor was more essential than the concept of LOR. Without NASA's adoption of this stubbornly held minority opinion, the United States might not have reached the moon by the end of the decade as President Kennedy had promised. Without LOR, possibly no one even now-near the beginning of the twenty-first century-would have stepped onto the moon."

Bell had submitted their LLRV concept to NASA in late 1961 and mirrored the NASA FRC concept that began work in May 1960. The NASA-funded $50,000 Bell LLRV feasibility study in January 1962 established the formation of the NASA FRC LLRV project office. I was scheduled to leave in August 1962 and had months before ordering a new 1962 Oldsmobile 4-door sedan to replace my worn-out 1957 Ford station wagon. For the drive to Buffalo, I received the car in June or July and then fabricated a system of bunk beds for my four sons: Greg (3 yrs.) and Bob (1 yr.), to sleep on the right half of the back seat, which was over the car's full-width bed, and cross-wise underneath the bunk bed for Allen (8 yrs.) and Gary (6 years). It worked well for long trips for two or three years.

We were all packed and ready to ship whatever household goods were allowed by the travel regulations. Then, in August, we hit a snag: Our contracts manager, Lloyd Walsh, who was leaving NASA Headquarters through the lobby with the $3.6 million contract to mail to Bell, was asked by a NASA aide if the contract had been mailed yet. Lloyd said no, but he was just about to drop it in the box outside. The aide insisted he not mail it, as a funding problem arose. This unexpected delay lasted till December 1962, so our trip was postponed until after New Year's 1963. When we finally arrived in Buffalo at 2 a.m. during a blizzard, we witnessed a vehicle that had just rolled over on its top, tires spinning, on the NY Thruway.

C. Wayne Ottinger

A Bad Day

Extending my time at FRC added one more memorable X-15 flight event. I was on the mission control console in NASA ONE during John McKay's November 9, 1962 flight. (from page 433 of X-15: Extending the Frontiers Of Flight. A NASA History Series)

"On November 9, 1962, Jack McKay launched X-15-2 from the NB-52B on his way to what was supposed to be a routine heating flight (2-31-52) to Mach 5.55 and 125,000 feet. Just after the X-15 separated, Bob Rushworth (NASA-1) asked McKay to check his throttle position, and McKay verified it was fully open. Unfortunately, the engine was only putting out about 35% power. In theory, the X-15 could have made a slow trip back to Rogers Dry Lake, but there was no way of knowing why the engine had decided to act up or whether it would continue to function for the entire trip. The low power setting seriously compounded the problems associated with energy management since the flight planners had calculated the normal decision times for an emergency landing at each of the intermediate lakebeds based on 100% thrust. The computer power at the time was such that there was no way to recompute those decision points in real-time, so the mission rules dictated that the pilot shut down the engine and make an emergency landing. McKay would have to land at Mud Lake.

As emergency landing sites went, Mud Lake was not a bad one, being about 5 miles in diameter and very smooth and hard. When Rushworth and McKay decided to land at Mud, the pilot immediately began preparing for the landing. The engine was shut down after 70.5 seconds, the airplane turned around, and as much propellant as possible was jettisoned. It was looking like a "routine" emergency until the X-15 wing flaps failed to operate. The resulting "hot" landing (257 knots) caused the left main landing skid to fail, and the left horizontal stabilizer and wing dug into the lakebed, resulting in the aircraft turning sideways and flipping upside down. Luckily, McKay realized he was going over and jettisoned the canopy just prior to rolling inverted. The unfortunate result was that the first thing to hit the lakebed was McKay's helmet.

As was the case for all X-15 flights, the Air Force had deployed a rescue crew and fire truck to the launch lake. Normally it was a dull and boring assignment, but on this day they earned their pay. The ground crew sped toward the X-15, but when they arrived less than a minute later, they found that their breathing masks were not protecting them from the fumes escaping from the broken airplane.

Fortunately, the pilot of the H-21 recovery helicopter noted the vapors from un-jettisoned anhydrous ammonia escaping from the wreck and maneuvered his helicopter so that his rotor downwash could disperse the fumes. The ground crew was able to dig a hole in the lakebed and extract McKay. By this time, the C-130 had arrived with the paramedics and additional rescue personnel. McKay was loaded on the C-130 and rushed to Edwards, and the ground crew tended to the damaged X-15. At this point the airplane had accumulated a total free flight time of 40 minutes and 32.2 seconds.

It had taken three years and 74 flights, but all of the emergency preparations had finally paid off. In this case, as for all flights, the Air Force had flown the rescue crew and fire truck to the launch lake before dawn in preparation for the flight. The helicopter had flown up at daybreak. The C-130 had returned to Edwards and carried another fire truck to an intermediate lake (they were possibly the most traveled fire trucks in the Air Force inventory). The C-130, loaded with a paramedic and sometimes a flight surgeon, then began a slow orbit midway between Mud Lake and Edwards, waiting. Outside the program, some had questioned the time and expense involved in keeping the lakebeds active and deploying the emergency crews for each mission. The flight program was beginning to seem so routine. Inside the program, nobody doubted the potential usefulness of the precautions. Because of the time and expense, Jack McKay was resting in the base hospital, seemingly alive and well. Had the ground crew not been there, the result might have been much different.

Although the post-flight report stated that the "pilot injuries were not serious," in reality Jack McKay had suffered several crushed vertebra that made him an inch shorter than when the flight had begun. Nevertheless, five weeks after his accident, McKay was in the control room as the NASA-1 for Bob White's last X-15 flight (3-12-22). McKay would go on to fly 22 more X-15 flights, but would ultimately retire from NASA because of lasting effects from this accident. X-15-2 had not fared any better, the damage was major, but not total. On November 15 1962, the Air Force and NASA appointed an accident board with Donald R. Bellman as chair. The board released its findings, which contained no surprises, in a detailed report distributed during December 1962. Six months after the Mud Lake accident, the Air Force awarded North American a contract to modify X-15-2 into an advanced configuration that eventually allowed the program to meet its original speed goal of 6,600 fps (Mach 6.5). Because of the basic airplane's ever-increasing weight, it had been unable to do this, by a small margin."

X-15 #2 after crash on Mud Lake, November 9, 1962

The next day I rode on the NASA R4D (DC-3) to the crash site at Mud Lake and marveled at the condition of the aircraft and the fact that Jack had survived. We held our breath in NASA One almost an hour after the crash, not knowing Jack's condition, even if he was removed from the cockpit. Not mentioned in the document above was the award from the White House given to the helicopter and ground rescue crew for their bravery and skill in rescuing Jack. Jack flew the X-15 for 22 more research flights again almost four years after this accident.

After work one day, I followed just behind Jack on the flight ramp to the NASA T-Bird, where he told me to go back, grab a chute, and get in the back seat for his check flight in the T-Bird. As I got in, he had a big grin and proceeded to climb out to Mt. Whitney, and with the wing tip pointed onto the top of the peak, pulled the highest "g" turn he could to get the best possible view of the mountain. He was a real personable guy.

In March 1964, I spent a month with his twin brother, Jim McKay, in Niagara Falls. Jim was the NASA structural engineer assigned to review the Bell design/test data before we took delivery of the two LLRVs. My family had returned to Lancaster to get the kids in school by then. So Jim and I would come on our off-time tour of the Niagara Falls area, including the two forts on opposite sides of the Niagara River, Fort Niagara on the American south side and Fort George on the Canadian north side. We spent time in Madame Tussaud's Wax Museum in Niagara Falls and kept busy during all our off-duty hours.

As a young child, Jim stepping on the live grenade buried outside his house, lost his right arm below the elbow, and his leg had to be amputated below the knee. That set the destiny of Jim as an aeronautical engineer, even though he owned his private airplane, while his twin brother, Jack, it was said, had more rocket time than any other test pilot. The shell was found on a hunting trip and placed on their mantel by their mother's servant; after she told him to get rid of it, he buried it in the background.

14 Months at Bell Aerosystems ON the LLRV Design Team
(Wheatfield Plant, Niagara Falls, NY)

NASA funding and bureaucratic delays for five months of getting the LLRV contract to Bell were putting pressure on the program. Paul Bikle was anxious to make this project's first contribution to NASA's space program a technical and financial management success. As with any advanced aircraft/technology program, fixed-price contracts are not suitable, as too many unknowns exist to assure profitability to the contractor, so cost-plus contracts with various forms of fee arrangements are part of negotiations. In Bell's case, they had just lost or completed some government programs, causing their overhead rates to rise significantly over their bid as the LLRV program progressed. My instruction from our LLRV program manager, Don Bellman, was to follow the technical developments for design and build and to watch the contract and business aspects of the program closely so Bikle knew we were on top of managing the program.

At 29, I had much to learn and was eager to do it. I spent my first month in an office at the DCAAS (the Defense Contract Audit Agency) in the building next to the Bell plant on the Niagara Falls Air Base. The Air Force Colonel in charge there was accommodating. The Bell engineering staff quickly arranged for me to have a private office in their plant to improve participation in their design activities. My specialty was propulsion, and I faced several significant design challenges as the propulsion system designs, rockets, and jet engine progressed. I was pleased that I could contribute to the team effort, expanding my knowledge from their considerable depth of engineering talent and learning a lot about structural, avionics, electrical, and systems integration, all with significant challenges to meet the mission requirements.

I also learned much about handling internal conflicts in a large company operating under stress in a downsizing mode amid projected financial overruns. The Navy's new PERT system of computerized reports, charts, and scheduling accompanied by cost management was required in the contract, and my Bell associates and I had learned both its good points and serious limitations, including the quality of the inputs from individuals responsible for defining and performing the tasks forming the database. The negative conclusions of the periodic reports caused friction between departments. I was in the middle as the same reports that reached the NASA FRC triggered serious contractual and management issues requiring innovative solutions and cooperation to keep the program going.

The first Bell LLRV program manager had to be replaced during these difficulties, and the replacement, retired Navy Commander John Mullen, made the rest of the program successful. I had good relations with Bell's top contracts, engineering, and manufacturing departments, which helped significantly as we negotiated the movement of significant portions of the LLRV assembly and testing work from Bell to NASA DFRC. These efforts allowed FRC Director Bikle to not ask for more funds from NASA Houston. During these times, I frequently needed private telecoms with Lloyd Walsh, the FRC contracts manager, and others at FRC in Dornberger's unoccupied office.

Ken Levin, the Bell LLRV Technical Director, was a super-talented engineer and a smart problem solver, whether technical or management issues. His assistant technical directors, John Ryken (stability, control, and performance) and Bill Jackson (propulsion), both matched Ken's skills. All were good models for my eventual role as the Lunar Landing Training Vehicle (LLTV) Technical Director in the follow-on program John Ryken led as Program Manager in 1966. In the winter of 1963, Ken Levin and I visited Wright Patterson AFB in Dayton, Ohio, to discuss our selection of the General Electric CF-700 aft-fan engine for the LLRV gimbaled engine that provided earth-g offset for lunar simulation. We chose to go by train due to weather conditions in Buffalo; it was quite a trip for a desert rat.

Changing the LLRV Configuration

(NASA Monograph SP-2004-4535, I co-authored)

During the spring of 1963, events beyond the LLRV's development prompted a change in the vehicle's fabrication. In a move resulting from the Apollo program's adoption of the Lunar Excursion Module (LEM) — later simply called the Lunar Module (LM) — Ken Levin recommended reducing the size of the LLRV and moving the cockpit so that it straddled the legs of the vehicle. This change, Levin argued, would reduce the height of the LLRV by about one-third and help solve problems related to weight, cg, and jet engine inlet distortion, all of which stemmed from the earlier placement of the cockpit cab over the jet engine. A few days later, on May 7, Levin elaborated on configuration changes in a meeting at the FRC, also attended by Dick Day from the MSC.

During these discussions, the advantages of Levin's proposal became apparent. The changes would make the LLRV more like the LEM, reduce its weight by at least 60 pounds, as the structure was less complicated, reduce engine inlet flow distortion, give the pilot better downward visibility, and increase payload flexibility. There were, admittedly, a few disadvantages. The changes would make the LLRV less symmetrical, for one. For another, implementing the structural redesign would require an additional 1,400 hours, delaying the production schedule by about two weeks. Despite these drawbacks, the group agreed, believing the advantages of the new design far outweighed the disadvantages. Fortunately, the changes entailed no amendments to the contract since both designs seemed to satisfy the contract's Statement of Work.

Bell was urged to proceed at once with the new design. However, concerned about the design changes' impact on the delivery schedule, the FRC asked Bell to provide more detailed production information so that the schedule and costs could be tracked monthly. Reluctantly, Bell representatives agreed to provide this information, even though doing so would add costs they believed were unwarranted. The Bell p assistant technical director for propulsion, Bill Jackson (formerly a manager for jet engine productions in Detroit,) lived close to us in Williamsville, NY. In the summer of 1963, we both took our families to Lynn, MA, near Boston, to camp out for a week and witness the 50-hour preliminary flight rating test of the GE CF-700 2V running on a vertical test stand at the GE factory for small jet engines.

A Sad Day for the Nation and Family Stress

I was sick with flu at home when the news broke about the assassination of John F. Kennedy in November of that year; a sad day for all. A weekend trip to DC a few days later was taken with our four small boys. Sondra was coping with her dad's religious issues then but did her best to fulfill her role as a mother, and we were all trying hard to make it work, not realizing the extent of the emotional stress building on her part. We hired a housekeeper to help winter of 1963, which gave her the emotional support she needed at the time (despite the longer-term adverse effects occurring 17 years later). During the winter blizzards of 1963, Sondra was at home in Williamsville, NY, in a wheelchair with a broken ankle, and I was at a late dinner with visiting NASA people from Edwards. Returning home, I discovered the garage door open and smelled of smoke. Earlier, I had neglected to empty the dryer's lint filter in our laundry facilities in the basement. The lint filter fire produced lots of smoke, and after the fire department extinguished it, they left the garage door open to clear it despite the sub-zero temperatures. No cell phones in those days, and I carried long-term guilt for the whole incident.

We bought season tickets to the 1963 summer Broadway shows close to Niagara Falls in Tonawanda alongside the river on the beautiful grass grounds of the Wurlitzer organ plant. In addition, we could take the visiting NASA FRC crew (which had stopped by the Bell plant after delivering the LLRV small-scale wind tunnel model to NASA Langley) to see Ethel Merman perform. Joe Walker piloted the Aero Commander, and the few engineers aboard could get a first-hand look at our LLRV design progress.

Christmas of 1963, some new neighbors had just moved in the week before from England: In partnership with the British Hovercraft Corporation (BHC), Bell had hired Wilfred J. Eggington for a senior engineering position working on air cushion vehicles. He and his family spent Christmas Eve with our family. Wilf held executive positions at Aerojet, Litton Industries, and Rohr Marine and was responsible years later for granting me consulting work. At Bell, we completed the manufacturing and fabrication portion of the LLRV contract, and I began planning and coordinating the flight test facilities out at FRC Edwards. At the LLRV Rollout at Bell in April 1964, I was 30 years old.

Not long after the assassination of President Kennedy, I made a trip to Long Island, NY, to visit the Grumman Corp, the designers of the Lunar Excursion Module for Apollo. I had not seen my first cousin Dale Minor since we were kids, and as he worked for a radio network in New York City, I visited his apartment on my way to Grumman. We played catch-up all night and parted at dawn. After Berkeley and his Korean War service, Dale continued his journalism career to cover racial conflicts in the South, Vietnam, and Israeli Wars. He later was Walter Cronkite's writer at CBS. Dale's1970 Book, *"The Information War,"* was published by Hawthorne Books. The book was on Walter Cronkite's office shelves when he joined him in 1982. Col. Jack Broughton's book *"Thud Ridge,"* published in 1969, dovetails partly with Dale's account for the same period of the Vietnam War.

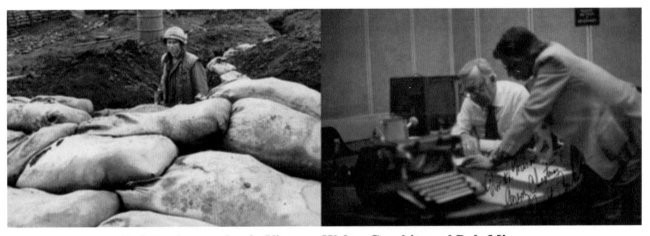

Dale in the trenches in Vietnam Walter Cronkite and Dale Minor

The Bell LLRV Technical Director hosted a party at his home in Buffalo on February 9, 1964. That Sunday evening, the Ed Sullivan Show hosted the first performance of the Beatles in the US.

The shift of work from Bell to FRC to reduce the contract overrun required taking about 20 Bell technicians and engineers from Niagara Falls to Edwards to work with us. First, we installed the electrical and rocket system plumbing. Then, we completed the final assembly work and systems checkouts, preferably in the hangar and then on the test ramps for propulsion systems testing. Bikle could cover many of the FRC costs under existing budgets, so the Bell overrun was not earning disfavor with Houston.

Bell held a press conference with local TV stations in the Buffalo region in early April 1964. Following the photograph taken at the rollout ceremony, the legs were removed for the LLRV #1 in preparation for transportation to Edwards in two air-cushioned (suspension system) moving vans. Two drivers were used in each van, and the trip took less than 50 hours. LLRV #2 was delivered by the same arrangement soon after but not assembled for testing until the flight research program was completed in late 1966.

Bell Vice President General Walter Dornberger (former commander of the Army portion of Nazi Germany's missile installation at Peenemünde), Wayne Ottinger (NASA LLRV Project Engineer), Bill Smith (Bell Vice President) at the LLRV Rollout in April 1964.

Due to limited assembly at Bell, a unique challenge met by NASA FRC was developing a sophisticated weight and balance measuring control system. VTOL (Vertical Takeoff and Landing) aircraft pilot handling qualities values historically indicated Grumman, the new NASA prime contractor for the Lunar Module (LM), had a dilemma. The control systems would likely require higher handling qualities for the much lighter LM Ascent Stage than the heavier Descent Stage. The NASA FRC years of aircraft control system expertise envisioned a major flight research objective to determine the acceptable VTOL handling qualities for an LLRV gimbaled jet flying in a lunar environment at five-sixth "G" and as though it was a vacuum.

The LLRV Design Keystone

Early exploration of the lower limits of control authority to report to Grumman's LM flight controls designers was a major contribution to the Apollo program. The research flights completed to get this data for Grumman were made with tight controls on aerodynamic limits for wind speeds throughout the flight envelope. In addition, early morning flights with calm and cooler temperatures for adequate engine thrust dictated takeoff requirements.

The Keystone was the Center of Gravity (CG.). Ray Manners (a Bell engineer with a distinct British accent) did a superb job working with our FRC engineers and technicians accomplishing this to the accuracy levels required for safe flight.

We limited all flight operations to less than five or ten knots of wind speed for our research flights. This low control authority adapted by Grumman allowed them to avoid serious design challenges of over-control on the ascent stage of the LM, which used the same flight control system for both the LM ascent and descent stages. In addition, the low control authorities proved acceptable (unexpected) by the LLRV research flying, allowing Grumman to avoid over-control as the ascent stage launched off the descent stage, returning from the lunar surface.

This sensitive cg (center of gravity) control started with a one-tenth-of-an-inch spherical accuracy for static flight-ready vehicle/engine assembly. However, the flight management challenges of ensuring the takeoff cg with the pilot aboard were within a quarter-of-an-inch spherical accuracy. Also, the subsequent in-flight cg did not exceed a one-half-inch sphere movement for a vehicle with 24 hydrogen peroxide thrusters and a jet engine burning JP-4 jet fuel. These requirements and the gimbaled jet vehicle design required a unique and complex weight and balance system to be developed, as shown below. An extensive procedures manual involving mathematic calculations yielded the system design.

President Johnson's Visit June 19, 1964

LLRV mounted on the weight and balance fixture

On June 19, 1964, just two months after the two LLRVs arrived at FRC partially assembled, Edwards AFB received a visit from President Johnson. NASA displayed the X-15 and the LLRV on the flight apron with an impressive lineup of fighters, bombers, transports, and helicopters. In addition, the new C-141 Starlifter transport carrier was displayed with troops and equipment representing a full payload stretched out behind the aircraft making the flight line an impressive sight.

The FRC deputy director, Dee Beeler, requested that I mount the LLRV on the Air Force flight line with the X-15 and display it in a way that demonstrated the gimbaled jet engine as though it was in a lunar sim mode. He also wanted me to add a chemical to the rocket system and blow smoke out of the sixteen-attitude and eight-lift rockets. However, I maintained that adding anything to the rocket system catalyst beds would void the Bell warranty on the H2O2 rockets. And Dee accepted my offer to insert red light bulbs into each rocket nozzle and have them light up, hopefully when the President would move the control stick and rudder pedals.

The Barry Goldwater campaign bumper stickers were labeled AUH2O, the chemical symbol for Gold and Water, and I quickly had a red warning light lens made with AUH2O engraved and made to slip in and replace our H2O2 Low Warning light. The test cart hooked up to run the rocket lights also had a flasher to operate the new warning light when Johnson was in the cockpit.

The day before the visit, I took Paul Bikle into the hangar and showed him the Goldwater warning light flashing. He had no idea what it meant. When I told him, he laughed and left. The following day, Dee Beeler called me and insisted I bring the lens to his office. I had to tell him Joe Walker had it and had already left for the flight line. The secret service foiled our whole plan, though, as they would not allow the President to sit in the cockpit even though the seat was disarmed.

Meanwhile, Joe saved my job, as he hadn't inserted the Goldwater warning light. The new red lights in the rockets worked when Joe wiggled the stick from the cockpit.

The LLRV display showing a pitch/roll gimbal demonstration for President Johnson's visit

Paul Bikle, Director FRC (L) Joe Walker, FRC Chief Pilot (Ctr.) President Johnson (R)

(L-R) Ken Levin, Bell LLRV Technical Director; Wayne Ottinger, NASA Project Engineer, Bill Giesel, Bell Aerospace President

Innovative Solutions to Avoid Risky VTOL Tether Tests

There was a bad history of crashes and fatalities in past VTOL (Vertical Takeoff and Land) aircraft tether testing. The risk would have been even more significant with the LLRV hybrid (rocket and jet fuel) propulsion system and complex first-ever fly-by-wire control system with no mechanical or hydraulic backup systems. During the extremely short seven-month assembly, system checkouts, and jet and rocket propulsion systems testing (April 1964 to October 1964), tether testing was avoided by two different test fixtures. They provided complete closed-loop hot firing of the lift and attitude control rocket systems (Reaction Controls), the jet engine, and the hydraulic gimbal control system for the jet engine stabilization and lunar simulation operations.

The first fixture provided the attitude rocket system in pitch and roll to be hot-fired. System checkout and pilot training were enabled by removing the pitch and roll gimbal actuators and providing soft bungee stops at each leg. Yaw was a non-critical mode, checked out on the ramp before takeoff. This closed-loop testing on a test fixture was critical to fine-tuning the analog electronic control system. I operated the LLRV using this fixture in the initial testing, and NASA trained all astronauts and test pilots in this manner.

LLRV on CG fixture for attitude rocket system testing

The second fixture was only used once and was designed to hot-test the lift and attitude rocket systems and the jet engine, including its hydraulic pump and gimbal control system (in two different configurations, one for pitch and one for roll). The procedure required all the pitch testing with the opposite or roll gimbal actuators, replaced with solid rods, as we tested the lift and attitude rockets and the jet engine running and operation of the pitch gimbal actuators. Once these tests satisfied the fine-tuning of the analog electronics in the pitch axis, we used a crane to lift the LLRV off the test fixture, rotated it 90 degrees to the roll test position, and repeated the same integrated testing.

The LLRV on the single-axis test fixture (roll position), Jet and Rocket System Integrated Testing

For the follow-on LLTV, the CG fixture testing was retained only for initial system checkout and subsequent pilot training. Initial LLRV propulsion tests were made for the rocket systems with no jet engine installed, then after the engine was installed, Joe Walker and Don Mallick made over 50 flights.

The Flight Research Program

Early LLRV flight with casters (replaced with footpads to prevent rolling)

U.S. Army Col. Jack Kluever joined Mallick to share the rest of the LLRV flights, a total of 198, ending June 30, 1966, with six more for LLRV #2 in early February 1967. Col. Kluever was on assignment to FRC from the Army as chief of their flight test facility at Edwards. On a Thursday, he had taken Joe Walker out for a helicopter check ride in 1963, where Joe, discovering the difficulty in hovering for the first time, took actions demonstrating his determination: The following Monday, Jack asked the boss, Joe Vensel, where Joe was. The answer came, "Oh! Joe's in Pensacola at the Naval Air Station in helicopter school." Joe received the Collier Trophy in 1961 for his record flights in the X-15 (also the DFC, Distinguished Flying Cross. for his WWII P-38 combat missions) and was determined to make the first LLRV flights as the project pilot. He was dedicated to sitting in the hot cockpit to operate the early jet engine and rocket system testing during the summer of 1964. I was challenged to bridge the gap between his lack of patience and tedious checklists as the ground crew, fully dressed in protective rubber suits, worked through our carefully planned operations. I had to convince him and the crew that we were doing our best to expedite operations while maintaining critical safety practices.

On flight #5 (11/25/1964), Walker was to make a demonstration flight for the national press, which consisted of three major TV networks and newspaper and Aviation Week reporters. Two or three large buses full of press people and TV crews arrived at the NASA FRC facility just north of the main base before dawn. Our flight operations, though, were at the old base eight miles south, where in those early days, we operated out of the aero club WWII hangar till we could get our own Butler building set up. Unfortunately, there was a storm the night of 11/24/64, with high winds and cold weather, as we reported at about 2 a.m. 11/25/64 for pre-flight and to load propellants for takeoff just after dawn. With the weather so bad, the crew gave me a tough time keeping at it, but I insisted. Finally, over coffee and rolls in the cafeteria, the Public Relations Officer, Ralph Jackson, told the press that we had to cancel due to weather. I was on the radio to Della Mae, the secretary in the Flight Ops Division, that we had Joe Walker in the cockpit, and they had better get on their way as we were going to fly.

As the busses all came around the curve approaching our ramp area, Joe took off in the LLRV as the busses arrived. The press unloaded from the buses, scrambling to get their cameras going, gawking at this crazy-looking VTOL aircraft flying over their head. Joe flew a few minutes demonstrating pitch and roll maneuvers around the gimbaled jet in the local vertical mode (always pointed vertically to the earth). Later use in lunar simulation mode allowed the vehicle to respond, simulating as though it was on the moon with 1/6-g gravity. Since we took advantage of the calm hole in the storm center passing over our flight apron at the perfect time, the crew never re-challenged me for our preparation for several demonstration flights.

Our first few flights were made using NASA One's mission control room; we had displays of the many telemetering parameters, our only instrumentation on the LLRV besides hardwired ground support equipment. Due to weight limits, no airborne film recorders (such as those used on the X-15) were used. After the first flight, we all agreed that we needed a mobile control van at the south base flight apron equipped with telemetering displays and communication equipment to allow the ground flight controller to be on site for ground and flight operations. NASA One would be back up and on the communications loop using the displays in their control room. The vehicle selected was a modified Ford Condor Motor Home built in El Monte, CA. Initially, a rack was installed on its roof to assist pilot removal from the cockpit. However, the vehicle was soon so overloaded with the equipment it required blocks to offset the springs' sag; The "mobile" part of the NASA 9 designation became a joke.

NASA 15 Mobile Control Van for South Base flight operations

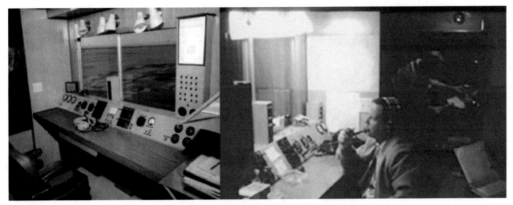

Van interior & me with pipe

A Brave Act Averts Disaster

The original design used teleflex cables to connect the throttle in the cockpit to the jet engine, which required routing over both the pitch and roll gimbals to connect to the engine connection. The cable friction passing through the loops over the gimbals (which moved 40 degrees in all directions) was too much for ideal operation by the pilots. So in the first couple of months of flying, we decided to replace the cables with a WWII fighter hydraulic throttle control one of our machinists, Charlie Lynn, had in a drawer.

We used it nicely for a few flights until one cold morning in mid-January 1965 when Don Mallick, who had already made a few flights, was ready for takeoff. The ground electrical power cart and engine (air) start cart were still all connected as Don started the jet engine. The RPM climbed quickly, and Don tried to bring it back down but had no control. Full of jet and rocket fuel, the vehicle rose on the landing gear struts but was still connected to the ground cart by the large electrical cable and the large air hose from the start cart. We quickly realized there was no control of the jet engine, which was up and running at somewhat below takeoff thrust. One of our crew, Bill Wilson, who dressed in his protective rubber suits, got his dykes and asked to go in and manually shut the engine down. Bill promptly entered the area next to engine with the hot exhaust blast reflecting off the cement apron a foot below the exhaust nozzle. Bill cut the safety wire on the main fuel control throttle lever and manually forced the lever to shut the engine down safely. We were fortunate that the RPM was low enough to avoid the vehicle from rising more and potentially tipping over in a disastrous fire with an out-of-control jet engine.

By the next day, we had determined the cause was that there was no temperature compensator in the system required to compensate for the rigging difference of the hydraulic system operating in large delta temperature environments (as we had with very cold mornings and warm afternoons). Don Bellman, our program manager, designed a temperature compensator for our system, which the machine shop made overnight, and after bench testing the system, we were back flying in less than a week.

Don Mallick, who flew 79 of the 198 LLRV research flights

View of The LLRV South Base Flight Test Complex

My hours were so varied no carpool was available, as I had on the X-15 for the seventy-mile round trips, so I bought a used 1959 Volvo. The LLRV crew was an innovative bunch of guys, good at their work, but liked to play jokes, including a good one on me: I used a NASA black & white station wagon to get back and forth from the south base to the main FRC facility's 4800 building. While driving past one of the crew's carpools that had just left one afternoon, I found my Volvo on top of the large flatbed truck and the forklift that had put it there. I whipped around and chased the crew before they got to the security gate, demanding the keys to the forklift, as I knew they had taken them to force me to take the government vehicle home. They, sure enough, had the keys, and I got my Volvo off for my trip home.

On a Saturday, we arrived to find our hangar had been broken into. Many of the tools belonging to the mechanics and technicians were stolen. However, my ire notably rose when I discovered our freezer was missing the luncheon steaks I used in a diet competition with Col. Jack Kluever. Later, the Air Force traced the theft to one of their AP guards.

Chief Test Pilot Joe Walker planned to leave the LLRV project to fly the XB-70 Mach 3 bomber coming to FRC as a research test vehicle. Jack was already assigned on loan to NASA from the Army and was a natural replacement to share the rest of the flight program with Don Mallick. Along with the usual aircraft and systems manuals, I taught Jack the systems. Jack would join us at 3 a.m. on most days as we aimed to fly right after dawn (being very marginal on weight limits; the lowest temperatures helped us get more thrust). Jack had to lose about 30 pounds to start flying within our weight limits, so we made a 10-week bet on meeting our weight loss goals, 30 pounds for Jack and 20 pounds for me. Jack loved sweets, and after discussing ejection seat improvements over dinner, he left the table briefly. He returned to find all the restaurant's desserts tempting him; a few bucks for me on our weekly weigh-in. Jack made his first LLRV flight on December 13, 1965. In April 2011, I included our weight loss competition in my eulogy at his memorial service in Las Vegas.

Walker Demo Flight for Over 500 People

On flight #5 (11/25/1964), Walker was to make a demonstration flight for the national press, which consisted of three major TV networks and newspaper and Aviation Week reporters. Two or three large buses full of press people and TV crews arrived at the NASA FRC facility just north of the main base before dawn. Our flight operations, though, were at the old base eight miles south, where in those early days, we operated out of the aero club WWII hangar till we could get our own Butler building set up. Unfortunately, there was a storm the night of 11/24/64, with high winds and cold weather, as we reported at about 2 a.m. 11/25/64 for pre-flight and to load propellants for takeoff just after dawn. With the weather so bad, the crew gave me a tough time keeping at it, but I insisted. Finally, over coffee and rolls in the cafeteria, the Public Relations Officer, Ralph Jackson, told the press that we had to cancel due to weather. I was on the radio to Della Mae, the secretary in the Flight Ops Division, that we had Joe Walker in the cockpit, and they had better get on their way as we were going to fly.

As the busses all came around the curve approaching our ramp area, Joe took off in the LLRV as the busses arrived. The press unloaded from the buses, scrambling to get their cameras going, gawking at this crazy-looking VTOL aircraft flying over their head. Joe flew a few minutes demonstrating pitch and roll maneuvers around the gimbaled jet in the local vertical mode (always pointed vertically to the earth). Later use in lunar simulation mode allowed the vehicle to respond, simulating as though it was on the moon with 1/6-g gravity. Since we took advantage of the calm hole in the storm center passing over our flight apron at the perfect time, the crew never re-challenged me for our preparation for several demonstration flights.

Our first few flights were made using NASA One's mission control room; we had displays of the many telemetering parameters, our only instrumentation on the LLRV besides hardwired ground support equipment. Due to weight limits, no airborne film recorders (such as those used on the X-15) were used. After the first flight, we all agreed that we needed a mobile control van at the south base flight apron equipped with telemetering displays and communication equipment to allow the ground flight controller to be on site for ground and flight operations. NASA One would be back up and on the communications loop using the displays in their control room. The vehicle selected was a modified Ford Condor Motor Home built in El Monte, CA. Initially, a rack was installed on its roof to assist pilot removal from the cockpit. However, the vehicle was soon so overloaded with the equipment it required blocks to offset the springs' sag; The "mobile" part of the NASA 9 designation became a joke.

pounds to start flying within our weight limits, so we made a 10-week bet on meeting our weight loss goals, 30 pounds for Jack and 20 pounds for me. Jack loved sweets, and after discussing ejection seat improvements over dinner, he left the table briefly. He returned to find all the restaurant's desserts tempting him; a few bucks for me on our weekly weigh-in. Jack made his first LLRV flight on December 13, 1965. In April 2011, I included our weight loss competition in my eulogy at his memorial service in Las Vegas.

Spidery-legged Lunar Landing Research Vehicle, with NASA test pilot Joe Walker at controls, demonstrates lunar approach techniques durin Edwards AFB test. Crowd gathered to watch

Space, Lunar Crafts Perform For Officials

EDWARDS AIR FORCE BASE, CALIF. (UPI)—The metal version of a wingless space re-entry craft designed to land like a conventional airplane will get its first test flight within the next three months.

Space agency officials announced this decision Friday after watching a flight of the triangular shaped M2 space craft and a test of the Lunar Landing Research Vehicle (LLRV).

Officials said the metal version of the M2 will mark another step in the development of a maneuverable space craft, and the LLRV test a step in the U.S. moon program.

The M2 will continue to be towed to 10,000-12,000 feet and released for a powerless gliding flight to a landing. More than 100 test flights have been made

The greying Thompson, a veteran of numerous X15 flights, quipped, "No, it's too dangerous."

The four-legged LLRV is being used by the federal space agency to develop techniques that will be used by astronauts in a similar craft in the first landing on the moon's surface.

After lifting the vehicle a couple hundred feet into the air, Walker eased the spidery LLRV to the ground with the centrally located jet engine counterbalancing five-sixths of the vehicle's weight. This downward thrust simulates the gravity on the moon which is one-sixth that of the earth.

Local Newspaper Article

Our demo flights had always been made at the South Base LLRV facility, but this demo required us to make the demo flight just behind the main FRC facility due to the high number of observers. NASA removed the tall light poles at the edge of the aircraft ramp to clear the area for Joe's demo flight with the entire crowd lined up along the south edge of the buildings. The NASA 9 mobile van (bent springs and all) was just in front of the group. For several weeks, Joe Walker had been stubborn and would not agree to move a critical lunar sim abort switch from the console to the top of the jet throttle for better accessibility. I was motivated to push him to do it, as a few months before, the Army had had a fatal crash of their XV-5A VTOL at Edwards, the root cause being the wrong location of a critical switch put in an awkward position on a panel just behind the pilot. The pilot lost his life in that crash, and I considered that the LLRV needed to recognize that and make the change before we had a similar problem.

Our lunar simulation controls were designed to weigh the vehicle in flight just before lunar simulation operations began. We used an accelerometer, a pressure transducer (rocket chamber pressure equivalent to thrust), and a gimbal position sensor. These provided the analog computer inputs required to calculate the weight for controlling the jet engine to 5/6 of that weight and then reduce it further for jet and rocket fuel burn-off. This early system configuration was sensitive to the rate the pilot would pull up on the lift rocket lift stick. If the pilot came on too fast with the lift stick, it would weigh the vehicle too light; the jet would be throttled back more than it should, and the rapid sink rate increase would require a lunar sim abort. Abort required the gimbals to be locked and the pilot to bring up the jet engine thrust to make a safe landing. This problem was experienced occasionally, and the required aborts were always successful.

Joe made a good takeoff and climb, stabilized at a few hundred feet, entered the lunar sim, and made a beautiful flight until about 50 feet above the ground when he had to abort out of the lunar sim. Though I knew why, the crowd never caught on, thinking it was a great flying machine and test pilot. After he landed, I climbed up the step ladder at the cockpit to shake his hand, and he blurted out, "Change the *** ***** switch!" I checked the data later and found that Joe had moved his hand from the jet throttle down to the lunar sim switch on the panel below a total of four times in about two seconds, approaching the limit of human reaction time for that maneuver. He was stubborn but smart to change after his disappointment in front of a large group .

My Move From Civil Service to a Government Contractor

For the LLTV, the training mission contrasted with flight research. It had a fixed flight profile, so we removed the in-flight weighing function so it wouldn't bother the astronauts. I had invited the Bell's Chief Engineer, Casey Forest, to join me in the control van for that demo flight. Not long after, he approached me to offer me the position of LLTV Technical Director. (I had been writing the LLTV's model specification for NASA Houston, incorporating their changes, including several actual LM (Lunar Module) components, for the trainers. My instinct was to get a grade promotion to GS 14 at FRC in the civil service and then another one to GS 15 when I would go to Houston to run the training there. Unfortunately, it took several months for NASA to decide I would only get one promotion when I transferred to Houston.

With the Bell offer making about a 25% increase in my salary and presented with the challenge to ensure the LLTVs would have a better chance of success with my presence there, I took the offer, starting there in June 1966. My last year at FRC required lots of travel to Houston and Bell as we were preparing for the astronaut training to begin as soon as possible after the research flying at Edwards. In addition, I was asked to produce the bulk of the model specification for procuring the new LLTV (based on the LLRV) and spent much of my time and travel on that task. In my absence, Bob Baron, who had started with me on the LLRV as his first NASA job, took over when I left and proved capable of managing the South Base flight operations.

It seemed the entire Apollo program was behind and accelerating to catch up. Many configuration changes were done expeditiously to feed the flight handling qualities defined for the training missions and the Grumman LM control system design. However, funding delays and bureaucratic obstacles were real impediments as well. NASA Houston provided additional engineering and technician support for our south base operation to formally document the LLRV # 1 modifications made in the two years of research flying. We used this support to build the LLRV #2 to be capable of flight status, where they had to come on with those rockets to kill the energy for a safe landing, and then only if there was enough fuel left. Col. Kluever and I took the issue on vigorously even though there was a lot of resistance to removing them. We argued that by removing them, we could use a small portion of the weight saved to upgrade to a newly-available, higher thrust ejection seat rocket which might save the astronauts in an accident. We won the argument, which may have saved the lives of Neil Armstrong, Joe Algranti, and Stu Present during their LLTV ejections.

Configuring the LLRV for Delivery to Houston

We housed the additional GE personnel responsible for updating the drawings and configuration control documents at the south base in two large office trailers next to the small Butler hangar (which contained the LLRVs). In addition, we retained Burt Adams, the Bell LLRV draftsman, to update the drawings. We were getting ready for the training ops, done at that time Ellington Field's NASA Flight Operations, later expanding efforts to Cape Canaveral. There were almost a hundred other NASA aircraft operating out of Ellington Field. NASA hoped to train all Apollo mission Commanders, LM Pilots, and their backups and planned to have three new LLTVs and two LLRVs supporting these sites. The NASA FRC crew and Col. Jack Kluever traveled to Ellington to provide the checkout and first LLRV flight demonstration. Col. Kluever, by then, was the Army's program manager in the Pentagon for the new Cheyenne helicopter.

FRC delivered LLRV #1 to NASA Houston/Ellington on December 13, 1966, and LLRV #2 on January 25, 1967. NASA requested that Jack Kluever make the first LLRV flight as a demo for the astronauts, and Jack flew the LLRV #1 (March 3, 1967) like the master he was, putting it through a remarkable VTOL demonstration that amazed the astronauts. I'm sorry I missed it. I was working hard at Bell on the LLTV. When I accepted the Bell offer, we leased our new Palmdale house to a Lockheed engineer on the FRC's recent research aircraft acquisition, the SR-71. We prepared to move into a new tri-level home, which backed up to a nice park and soccer field and was ideal for our kids and dog.

Saturday, June 8, 1966, was the tragic and fatal mid-air collision of Joe Walker's F-104 chase airplane into an XB-70. The accident occurred during a photo mission of several GE-powered aircraft flying in close formation with the XB-70, and Joe Walker and the XB-70 copilot, Major Carl Cross, were killed. I reported to Bell the following Monday, June 10, receiving a call from Edwards of the terrible news. I respected Joe and immediately remembered our families sharing memories at a Harlem Globetrotter's game in Palmdale.

My new office was part of the row of offices for technical directors in other Bell programs, including the LM ascent engine, the Agena Rocket, the Rocket, and Jet Belts, and in cooperation with British Hovercraft Corporation, a large Air Cushion Vehicle. In addition, the NASA RFP (Request For Proposal) had just been released as a sole source for Bell to purchase three LLTVs. Since I had written the model spec for that RFP as a NASA employee, the Houston contracts office called a meeting at NASA headquarters protesting my participation in the LLTV project. Lloyd Walsh, head of procurement and contracts at FRC, was in the meeting and informed Houston that under the tight schedule the Request For Proposal (RFP) was at least six months late) and critical nature of the Apollo training, they had little choice but to leave me in. And though Houston barred me from the negotiations, Bell used their privilege of getting granted time for a private caucus during the negotiations.

So for two weeks, I sat in a private conference room at the hotel across the street from the NASA center with a private telephone conference line. I had all the LLTV drawing files and reports next to me. I became known as "Mr. X". NASA had good reason to be upset. Their estimate for the contract was a couple of million under our bid. So they decided (foolishly, as it turned out) that they could save significant dollars by moving all testing to Ellington Field. Our job now was to make all the design changes and get the testing done before astronaut training could begin.

New components (a number in the rocket system) had to pass new component qualification tests. The avionics for the LLTV required a new automatic test cart. Also, a re-design of the flight controls to better correspond to the final LM control system was added to the work. The pre-flight testing for the LLRV had been tedious and done on the flight line using patch panels and manual techniques. Additionally, the propellant and jet fuel loading, performed by fully-suited ground crews (protection from the 90% H_2O_2 rocket fuel), meant extended pre-flight checkouts. The new test cart for the LLTV proved to be a schedule killer, too.

Earlier in the LLRV program, a senior Apollo manager (Air Force General Phillips) told the President of Bell Aerosystems, Bill Gisel, that if the LLTVs were unavailable before the moon landing attempts, they would go without the free-flight training. As a result, Gisel spent a million or more on long-lead tank forgings ten months before any NASA contractual coverage was provided. Bell was furnishing the LM Ascent engine and had a significant stake in the success of Apollo. As it turned out, according to several Apollo Commanders who landed on the moon, "the LLTV was responsible in large part for the six successful (6) safe landings and crew returns in the Apollo program."

On my first trip to FRC in the early summer of 1966, I took Alan King, a British flight test engineer Bell had assigned to help me with LLTV ground and flight testing. When we arrived at FRC, we had quite an introduction in the pilots' office as I introduced Bill Dana, an X-15 pilot and old carpool buddy, to Allen King. There was quite a laugh, as both names were those of popular comedians.

Soon after that trip, Neil Armstrong and I visited FRC to witness the ejection seat qualification test, which required a zero-zero ejection of the upgraded seat with a more powerful rocket. The new ejection seats would be installed in both LLRVs and LLTVs. As we watched the test, the seat/dummy assembly rose several hundred feet above the ground, appearing to be gyrating wildly on a disturbing trajectory. When the dummy landed, Neil, without a word, got quickly in his NASA station wagon and drove off back to the hangar. We later assured him the test was a success, as we discovered the dummy's boot had struck the simulated instrument panel when the rocket catapult powered the seat up the cockpit rails. Furthermore, we determined the test fixture caused the problem and was non-existent in the vehicles' cockpits, and the gyrations observed were a result of the outside force exerted on the seat as it left the test fixture.

The picture of the dummy, left face down in the desert and with all the observers leaving the scene with their backs turned on the dummy, inspired me to present that photo to Col. Jack Kluever in a meeting at the Weber offices in Burbank. The caption on the photo read, "Oh well. He was just an Army pilot." At a dinner for our NASA FRC Flight Operations Director, Joe Vensel, Jack's wife (having seen the photo with the caption), gave me a friendly threat with a dinner knife, but all was forgiven.

Soon after Neil's Gemini 8 mission, I was visiting Neil in his Houston office when I received a call indicating my 12-year-old son, Allen, was in the hospital with a concussion he received in a school boxing match. After the call, Neil reached over, got one of his Gemini 8 pictures, signed it, and wrote a message to Allen to get well soon. Neil was Commander of Gemini 8, and on March 16, 1966, he made an extremely difficult recovery. Finding the correct switch to turn off the Gemini capsule's stuck rocket system, he separated from the Agena rocket (whose mission called for the first docking of any two spacecraft). He prevented the Gemini capsule from spinning out of control. A few months later, on a visit to Bell Ken Levin, his wife, Sondra, and I took Neil out to dinner at the Seagram's Tower, which had a newly-opened rotating restaurant and a 360-degree view of the Niagara Horseshoe Falls in Canada.

That night, Neil told us the story of the Gemini 8 emergency. They had burned too much rocket fuel to reach the planned site in the Atlantic. Indicating that under the centrifugal forces, the spinning was so severe that it was too painful to turn his head; he and Dave Scott could only rotate their eyeballs to search for the panel displays and controls. He told of their landing in the Pacific, on the high seas waiting for the USS Leonard F. Mason to reach the capsule. They were in the capsule, bobbing in the water with three pararesueres (frogmen) riding outside the capsule with the attached flotation gear. The frogmen had parachuted from the Air Force aircraft dispatched from Okinawa and circled the spot as the main chutes brought the capsule down to the ocean. Neil said that as the men waited for the carrier, they became seasick; one of the frogmen took a few days to get over it. The sea state appearance in the posted videos from NASA does not match my memory of Neil's story that the swells were over 20 feet. So I checked with Dave Scott, his Gemini 8 pilot, who verified my memory of Neil's 1966 story. *"Wayne: You have a better memory than NASA-of-today (not surprisingly). We were picked up by the destroyer USS Mason approximately 3 hours after splashdown; and the problem with the sea state was the unreported 20 foot (about) swells. Perhaps the best source of the story can be found at: Official Gemini 8 Report Keep at it with your memoirs, you have many good stories to tell...... Best, Dave"* Neil also told us about the psychological impact of the new Saturn launch pad (which rose to the tremendous height of a 30-story building) as the crews approached the elevator walkway at the foot of the pad. As a result, they erected a cover over the walkway to reduce an effect Dave termed "Scary."

Through December 1967, NASA management (Dean Grimm, the Houston LLRV/LLTV project manager, his boss, Warren North, head of the NASA MSC Flight Crew Support Division, Deke Slayton, Head of the Astronaut Office, and Joe Algranti, Chief of the Aircraft Operations at Ellington) visited our LLTV project at Bell at least monthly. Our meetings were held all day and evening, ending with late dinners at a famous Italian restaurant in Niagara Falls. Algranti and Slayton landed in their T-38 on the Air Force Niagara Falls Airport runway adjacent to the Bell plant, while Dean and Warren usually came commercial. John Ryken, the Bell LLTV Program Manager who helped me get the LLRV ready for flight, chaired these conferences, and I was on the hook to account for either the progress or lack of progress on the detail milestones required to deliver three LLTVs.

These meetings followed a standard pattern of project reviews aimed at completing all outstanding action items, particularly those delaying delivery of the training vehicles. But the most significant delay in deliveries to Houston came unexpectedly.

Apollo 1 Fire
(January 27, 1967)

The aluminum structure shown in the following photo is a typical LLRV or LLTV leg assembly, designed with hydraulic struts for the landing gear to carry landing loads; it represents decades of Bell Aerosytems' structural design expertise. Bell's LLRV design provided ample safety margins, as minor flaws identified by conventional X-Ray quality control inspections were controversial. As a result, Bell had three LLTVs on their shipping dock ready for delivery to NASA in Houston in August and September of 1967. However, the then-recent Apollo disaster, in which the three astronauts were fatally trapped in the capsule fire during a checkout, resulted in NASA's Quality Control exercising veto power over the Bell engineering traditions. As a result, NASA's inspectors disputed the weld inspection specifications applications to small cracks revealed in X-Rays, and a two-month delay was experienced in the already-late training operations.

Bell's Executive Vice President, U.S. Army Brigadier General (Ret.) Richard M. Hurst had turned over the Redstone Arsenal to Werner von Braun and NASA. For about two months, I worked directly with General Hurst daily to develop the strategy on solutions to the NASA contractual hold on LLTV deliveries. In addition, we prepared for legal measures and hired ALCOA aluminum welding experts to help us convince NASA we were on good technical grounds for our position. We subjected identically processed joints to tests at much more severe conditions than shown to be reasonable, thus solving the problem. My most formidable management and legal challenge as the Bell LLTV Technical Director was preventing a further delay than the two months already imposed.

Typical leg structural joint

Once we solved the LLTV delivery two-month delays, I began recruiting Bell engineering and technical staff. We were relocating them to Houston for the highly-underestimated term of six months. NASA's estimate for the Bell support requirements for ground and flight testing and negotiated arrangements for temporary housing and rental of furniture. By December 1967, our family was settling into the new rental house in Webster, TX, near the NASA MSC and Ellington facilities. It included a backyard dock in a canal just off Clear Creek, extending west into Friendswood, east into Clear Lake, and then into Galveston Bay.

We started with about fifty families. An administrator assisted me in getting the people relocated and setting up payroll and local office functions for them. A major housing challenge was the 1967 Houston culture that disturbingly showed racial bias. Our avionics engineer, Bill Bascom (in charge of the LLTV avionics group that had developed the LLRV system), was a highly-talented black mechanical engineer. All efforts failed to find anyone to rent him and his large family temporary housing. Finally, recognizing the dilemma, Bell purchased a house for them in their chosen neighborhood, which they stayed in for over two years.

I worked to develop the extensive facilities for LLTV testing that would be integrated with NASA's early LLRV flight operations. NASA's contractor Dyna Electron furnished mostly technicians, mechanics, and inspectors. Their large fleet of around 100 aircraft, including about 50 T-38 jets for astronaut training. This required extensive coordination with the NASA Ellington Flight Operations and the office of Dean Grimm, supported by Jim Bigham, in the Flight Crew Support Division with the NASA Ellington Flight Operations. Our LLTV operations required extra office space and expansion of the LLRV flight hangar facilities at the end of a runway at Ellington Field. Several months were lost getting many subcontractors to complete tasks not recognized by NASA until we arrived on site. Then the usual technical problems on the LLTVs began causing delays as system checkout revealed hardware and software turned out to be over 110 personnel staying for over two years (compared to NASA's plan for 50 people for six months).

Our LLTV rocket system testing on the Center of Gravity (CG) fixture was a real schedule problem considering the need to not operate in the rain, which is frequently a problem. So we fabricated a tall metal shed to roll over the vehicle installed on the CG fixture. On our first hot firing test on a rainy night, shelter in place, a NASA rumor mill took off the following day. Our test engineer in the cockpit operating the LLTV was Tom Stafford (no relation to the astronaut Tom Stafford). That night, as the attitude rockets fired during the test, the hydrogen peroxide steam collected on the underside of the metal shelter covering; small droplets of raw peroxide dripped down on the vehicle, causing (to be expected) small fires of anything combustible under the canopy. The idea of saving schedules by operating in the rain went up in flames (literally), so we adopted the LLRV example of waiting for dry weather. Of course, the NASA rumor spread like wildfire the following day *"Tom Stafford, the astronaut, had flown the LLTV through the shelter, setting it on fire."*

A Few Breaks in the Grind

We could go to a Houston Astros baseball game a few times as guests in the Astrodome corporate suite, owned by the furniture rental company that supplied the crew from Buffalo. I purchased a sixteen-foot ski boat with a lap-strake wooden hull from our flight test engineers (who had just bought a new rig) and tied it to my dock by my backyard, just off Clear Creek. After work, two of my Bell colleagues from Niagara Falls, John Ryken, and Walt Rusnak, still in their business suits and ties, headed towards Friendswood, a town a few miles inland. The creek was subject to the change in tides off Galveston Bay, and this ride was at high tide, covering a hidden tree stump in the middle of the creek. We were proceeding at a slow speed, not even rising up on the hull as speed increases (planning). The hidden tree branch threaded a foot-wide hole near the back of the hull, and as the sudden stop threw me against the windshield, I saw stars momentarily. The hull hung up on the tree limb, keeping the boat from sinking, and only an inch or two of the windshield showing above water. Still, I threw a rope over the tree-lined shore in my shorts and swam to a dense thicket of thorny bushes. I told John and Walt I would try to get help and took off through the thorny trees and snake-filled high grass for several hundred yards to get to the road. It was almost dark by this time, but I could hitchhike into Friendswood, finding a creekside home with a boathouse and boat ready to launch. The owner was gracious enough to set out several miles back to the creek only to find John and Walt nowhere to be seen. Finally, home, I called their hotel and found out they had stripped down, rolled up, and thrown their business suits to the shore, run through the snake-filled grass, and hitchhiked home. The repair bill was less than $100 at the Houston repair shop. My reputation, though, as a host for after-work boat rides down the creek with water moccasins and prehistoric-looking large birds suffered a great loss.

During the summer of 1967, my parents and Nephew, Kent Olson, visited from Phoenix. Through the City of Montreal, I had reserved four days at a French condo near the end of the new Montreal subway built for EXPO '67. All nine of us took our new Chevy station wagon and drove to Montreal, acclimating to the French neighborhood and the new subway commute. EXPO '67 was a showcase for both the Soviet Union and the U.S. to show off their space hardware and defense weapons and was also where the president of France, Charles De Gaulle, gave a controversial speech.

Expo '67 in Montreal

Christmas Party in Our Backyard

There were 60 or 70 families from the Buffalo area on our Bell crew, with probably 50 or more kids recently arriving on the LLTV project site, so I decided a Christmas party should be thrown for them as a "welcome to Texas" gesture. In Lancaster, my two oldest sons, Gary and Allen, then 11 and 13, had attended preschool taught by Don Bellman's wife. A few years older than ours, their sons entertained children with their creative puppet shows, so Allen and Gary took on duplicating, as best they could, what they remembered from Lancaster and set up a puppet show for the party in our backyard. But the star of the show was Santa Claus in full costume and with bags of toys, arriving in a 30-foot fishing trawler at our backyard dock just off Clear Creek. Willy (Bill) Wilson (the LLR mechanic who shut down the uncontrollable jet engine for Don Mallick at Edwards) was the Santa Claus. He had been promoted to crew chief, and I had hired him to come to Bell and serve as maintenance chief for the LLTVs at Ellington. He had just purchased the fishing trawler, surprising his wife with it at the party. Little did we know how much a party was needed to get us through the next two years.

The First Super Bowl

On January 15, 1967, the NFL Green Bay Packers smashed the AFL Kansas City Chiefs, 35-10, in the first-ever AFL-NFL World Championship, later known as Super Bowl I, at Memorial Coliseum in Los Angeles. John Matson, the marketing manager for the LLRV/LLTV ejection seat manufacturer (Weber Aircraft), acquired tickets and hosted Bell and NASA personnel to attend the game and a post-game party at his home in Glendale. Leaving Buffalo on a Friday evening before visiting my parents in Phoenix, I flew to Los Angeles for the game on Sunday morning. My LLTV quality control engineer was already there and was to pick me up at LAX for the game and more business the following week. The airline captain gave the passengers a vote to either land in

Ontario or circle till the fog cleared. We finally landed at LAX and scrambled to get to the Coliseum late in the second quarter. Of course, we missed John and the others with the tickets, but luckily, I found the right ticket manager to take us to the third row on the 50-yard line. As we sat down, the halftime ceremonies started with a great shock: One of my Bell associates who lived in my neighborhood was half the two-man team starting the halftime show. They each started at the goalposts at each end of the field. They flew past each other at the 50-yard line, the noise of the hydrogen peroxide rockets echoing throughout the Coliseum. Wendell F. Moore (the rocket belt project engineer) had his office next to mine at the Bell plant, and his rocket belt's appearance at the Super Bowl was a secret well-kept. The "Hip Pack," a lightweight aluminum structure to transfer the rocket thrust into the hips for the rocket belt.

The First Ejection, May 8, 1967

During Neil's LLRV #1 flight at Ellington, the rocket fuel and helium (used to expel the peroxide fuel from the tanks) had unknowingly become exhausted. Our Bell rocket engineer advised the NASA engineer to communicate instructions to the pilot in real-time during flights. At Edwards, we always had a manual prediction, never depending on the unreliable "Low H_2O_2" warning system to shut off the lift rocket system. Still, his input to the NASA controller was never transmitted to Neil, resulting in the ejection and subsequent loss of the vehicle. NASA blamed the accident on the warning system but recognized that safe ejection was a critical and positive contribution to the LLTV program. A truth repeated two more times. NASA never modified the warning system they chose to blame for the accident.

Neil Armstrong's Ejection from LLRV #1, Ellington Field, Houston, TX, May 8, 1968

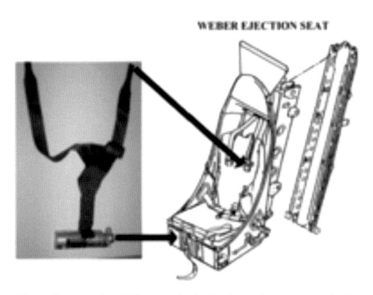

Seat-Man Separator (Butt Snapper): A Pyrotechnic Rotary Actuator winds up the pilot's harness he sits on, kicking him out of the seat as it reaches maximum altitude after rocket burnout.

Probably a half-hour after Neil's ejection, I sat next to him at a post-flight briefing. Much to his surprise, he began lisping his speech as he continued to talk. It turned out he was just discovering he had bit his tongue during his ejection rocket ride. It was swelling up on him, a tolerable result compared to the alternatives. A little later, as the crane was lifting the burnt wreckage from the grassy field, a half-burned snake crawled out from under the vehicle, no doubt saying, "What the heck was that?

This Gemini 8 3-axis side-arm attitude controller was modified and used in the LLRV, starting with flight #86 in March 1966.

The LLRV control system components, such as resolvers and potentiometers, replaced the like components in the Gemini configuration.

LLTV #1 ejection
by Joe Algranti, December 8, 1968

This flight was to be the last checkout of the LLTV before Neil Armstrong was to start training flights the next day. Under the hectic operational culture in Houston (with the large fleet of T-38s and a total aircraft inventory approaching 100 aircraft), the LLTV was allowed to takeoff that Sunday with an undetected wind shear condition was far exceeding the control capability. As Joe entered the wind shear several hundred feet from the ground, he lost control, and the LLTV rolled 90 degrees to the right. Fast approaching a 120-feet-per-second sink rate, he commanded a roll left to recover but rolled uncontrollably 90 degrees too far. Luckily the

LLTV, still out of control, rolled back to straight and level. Joe's safe ejection was even more amazing at that moment, three-tenths of a second before impact on the ground: the ejection seat performed far exceeding the expected capability. Though uninjured, Joe never flew the LLTV again.

The loss of training vehicles from the two ejections severely impacted NASA's training schedules and caused Neil only to receive six LLTV flights before launch on Apollo 11 (July 16, 1969). Armstrong had 21 LLRV flights from March 23, 1967, to May 8, 1968, all at Houston (without the Apollo flight hardware). After the two ejections, NASA expanded their mission control personnel by more than twenty, and operating procedures were improved so that Apollo landing mission commanders and their backups all received the LLTV training.

The Third Ejections
(Stu Present, January 31, 1970)

This dramatic 16-mm still photo shows the ejection rocket flaming out the jet engine. Sadly, Stu was killed the following year in a T-38 accident.

LLTV #2 ejection by Stu Present, 1/31/71

The accident investigation took less than one hour to reveal what caused the loss of the flight control system requiring Stu to eject. The root cause was the new "improved" General Electric DC generator retaining the residual magnetic field much longer during its failure, which disabled the emergency switchover circuit. In addition, the circuit was designed to sense voltage loss on the primary electrical bus (thereby enabling the emergency electrical bus). So why did it take less than one hour to discover the cause of the loss of the flight control system? The analog fly-by-wire system came alive after the ejection and the pilotless vehicle's rapid descent without the jet engine running. The switchover circuit finally enabled the emergency bus, and the battery allowed the flight control system to correct the vehicle's attitude for a normal landing. But without the jet engine running, the landing far exceeded the structural design limit, causing it to crash and burn. Though I had left the program two years earlier, I was gratified that despite losing three of the four vehicles, we'd retained all our test pilots alive and well.

NASA and Bell worked hard to achieve NASA's 1967 plan to add LLTV training to Cape Kennedy so the astronauts could get a training flight the day before launch. This plan involved establishing significant facilities at the Cape and required using the two LLRVs and the three LLTVs to support training between both sites. However, due to the loss of the first two LLTVs, this ideal plan had to be scrapped: The decision was made to abort using LLRV #2 and go directly to the LLTVs with real Apollo hardware and better reliability. So all training had to be done in Houston, and for the last 23 months of Apollo, only one LLTV was available.

Life Magazine asked to include the LLRV in their special Sept. '64 Apollo Simulator issue. So, Life was invited to the LLRV Design Engineering Inspection attended by representatives from several other NASA centers. Life's photographer, Ralph Morse, arranged strobe lights around the vehicle with the afternoon sun in the background, each pilot taking turns in the cockpit for their unique photographs. A year after Joe Walker was killed (in the XB-70 mid-air collision flying the F-104), Ralph Morse and I accidentally met for breakfast at adjacent tables in the hotel across from NASA Houston. I asked him if I could get a print of a picture he took that day in 1964. Ralph sent me the one of Joe Walker, much to my surprise and delight.

My cherished pictures of Neil Armstrong (bottom) and Joe Walker (top) in the cockpit of the LLRV

Life Magazine Photo, August 1964, Neil Armstrong in Cockpit

500 lb. rockets on each side of LLRV, (1) Normal Lift Rocket for lunar simulation mode, and (3) emergency lift rockets on each side of the vehicle mounted symatricsally around the vehicle picth axis

LLRV #1 crew photo, 1965. I'm on the right

LLRV #1 crew photo with GE support (to Build #2 LLRV), 1966. I'm second from the right

Yaw rockets purging with gaseous nitrogen (N2) after a hot firing

Rewarded Well by Bell In September 1968

Years of long hours and marital stress brought me to resign as the Bell LLTV Test Base Manager. After a three-month medical leave, I agreed to join Bell's Los Angeles sales office., I worked for Vice-President, Western Region USAF General John Mel Schweizer, Jr. We had moved back into our home at the Palmdale Country Club during the leave, and I prepared to return to work.

The Annual (September) the SETP Symposiums

Sondra and I attended most of the annual Society of Experimental Test Pilots (SETP) annual symposiums, traditionally held in the Beverly Hilton Hotel in Beverly Hills. After the technical sessions, participants would host hospitality suites. In September 1968, in one of these suites with probably a dozen other people, we heard Jim Lovell talk about the early planning meeting at NASA headquarters for what the crew would do for the Apollo 8 1968 Christmas Eve TV broadcast back to Earth. Lovell reported that one of the first brainstorm ideas to be blurted out was, "let's do a urine dump at Sunset. It is so beautiful." Of course, the final selection was the crew reading the first ten verses from the Book of Genesis by the crew. William Anders captured the historic Earth Rise photograph. The crew: Commander Frank Borman, Command Module Pilot James Lovell, and Lunar Module Pilot William Anders.

On January 27, 1967, the Apollo I fire fatal to the three crew members forced NASA to redesign the Command Module significantly. The Apollo 7 launch in October 1968 tested the new command module, followed by Apollo 8 with the Saturn IB with no Lunar Module. The Apollo 9 Earth Orbital Mission in March 1969 tested the final Saturn V booster for the first time. Finally, Apollo 10, launching on May 18, 1969, with the service module Charlie Brown de-docking Snoopy, the Lunar Module flew within 47,000 feet of the lunar surface, photographing the Apollo 11 landing site at the Sea of Tranquility.

Less than 31 months from Apollo I's fatal fire to Apollo 11's landing provide a legacy Americans should be proud of, particularly for the safe return of twelve moon walkers and three Apollo13 astronauts safe return after the accidental explosion en route to the moon.

Rewarded Well by Bell In September 1968

Years of long hours and marital stress brought me to resign as the Bell LLTV Test Base Manager. After a three-month medical leave, I agreed to join Bell's Los Angeles sales office., I worked for Vice-President, Western Region USAF General John Mel Schweizer, Jr. We had moved back into our home at the Palmdale Country Club during the leave, and I prepared to return to work.

<u>Neil Armstrong's Last LLTV Landing June 16, 1969, One month before Apollo 11 Launch</u>

A YouTube Silent Film of his seven minutes training flight, to eleven minutes for the press conference. The rest of the 26 minutes is repeated.

<u>Neil's last LLTV Training Flight</u>

<u>https://www.google.com/search?sxsrf=AJOqlzVo3WeRpBqxpoTSEEBlpgd5yWDipA:167 6582075914&q=Neil+Armstrong+LLTV+last+landing+My+16,1969&spell=1&sa=X&ve d=2ahUKEwiH-8j4-pr9AhX0ITQIHegVAaUQBSgAegQICBAB&biw=1408&bih=764& dpr=1#fpstate=ive&vld=cid:e44fb966,vid:iRJmuYInKcA</u>

My Two Bosses in the Bell Western Region Office (1969)

General Schweizer and General Wilson (on board to take over after Schweizer's retirement) were an interesting pair: Schweizer, as a 1st Lt. in the Army Air Corps in 1932, was sent to China to assist General Chiang Kai-shek in starting up his air force. He told us stories about assassination attempts upon his buddy and him as they traveled by riverboat. Before joining Bell, Schweizer served as the U.S. Air Force NATO representative in Paris and was fluent in French, as I found out later.

Del Wilson was somewhat younger and more conservative, and the office staff found it amusing when each General took it upon themselves to call and order the water cooler replacement. Arriving at the same time at our sixth-floor office, looking out at the LAX north runway, the vendor delivered two simultaneous orders, one water cooler full of bells and whistles, the other bare bones. No one had to guess who had ordered each of the units.

L.A. Times, OBIT: John M. Schweizer Jr.;
"Retired Air Force Brigadier General, Executive, January 26, 1996.

John Mel Schweizer, Jr., 89, retired Air Force brigadier general who moved from pilot to executive in military intelligence and then in commercial aviation. A native of Los Angeles, Schweizer studied at Occidental College, where he was a quarterback of the football team and lead trumpet for the Earl Bernett Orchestra. As a pilot in the Army Air Corps in the 1930s, he trained Nationalist Chinese military pilots.

He worked as a commercial pilot for American Airlines and Humble Oil until he was recalled to active duty in World War II. After becoming a brigadier general in 1953, Schweizer served as Air Force director of intelligence in Europe and later as director of intelligence for the Allied powers in Europe. The United States, France, Italy, and Greece honored him for his service. Schweizer retired from the Air Force in 1959 to become the European representative for Bell Aerosystems Corp. in Paris and later Bell vice president in Los Angeles.

In 1971, he became executive director of the Voices in Vital America, supporting the recovery of MIAs and POWs in Vietnam. [He died] on Monday in Camarillo of lung cancer.

USAF Major General Delmar E. Wilson
(Commander, Morón Air Base, Spain, Loss of 4 Hydrogen Bombs, January 17, 1966)

In November 1983, during the joint Spanish/American military exercise CRISEX 83, USAF B-52 bombers were allowed once again to enter Spanish air space and land at Morón Air Base. The B-52 bombers were previously banned from entering Spanish air space after the January 17, 1966 incident near Palomares, when an in-air refueling B-52G (s/n 58-0256) collided with a United States Air Force KC-135A jet tanker (s/n 61-0273). Two hydrogen bombs ruptured, dispersing radioactive particles over nearby farms. An intact bomb landed near Palomares. The fourth bomb was lost at sea, 12 miles (20 km) off the coast.

A search involving three months and 12,000 men was required to recover the device, however, despite the deployment of highly sophisticated technical equipment by the U.S. Navy; it was a local Spanish fisherman who finally guided them to find and recover the bomb.

In April 1966, the 16th Air Force was transferred from SAC to the USAFE, with USAFE taking control of the Spanish air bases at Zaragoza and Morón. Under USAFE, the Spanish bases became host to a growing number of deployments from CONUS. Morón received regular visits from Lockheed F-104C Starfighters of the 479th TFW from George AFB, CA. During the Cuban Missile Crisis, a squadron of F-104Cs was stationed at Morón. Concern at the height of the crisis led to these aircraft being transferred to Hahn Air Base in West Germany, where they strengthened the air defense of central Europe. Some-time later, when the crisis had passed, the aircraft returned to the U.S. via Morón. On April 1, 1963, their place was taken by F-105D "Thunderchief" fighter-bombers from the 4th TFW at Seymour Johnson AFB, North Carolina.

U.S. Combat ACVs for Operation in Rivers in Vietnam

Bell had modified (six) 25-ton British Hovercraft amphibious air cushion vehicles (ACV) for combat in Vietnam, three for the U.S. Navy, and three for the U.S. Army. General Schweizer assigned me the commercialization of ACVs for applications such as near-airport emergency rescue and fire-fighting operations, similar to those in

Vancouver, Canada. I was also assigned to the electronic products lines and systems work, supporting Edwin E. Gorin at Bell's AZ Operations (Tucson) Services. These included engineering, testing, and evaluating communications-electronics systems for the U.S. Army Electronic Proving Ground in Fort Huachuca.

U.S. Navy SK-5 Combat ACV

I received almost daily TWX reports from DoD on the ACV combat experiences in Vietnam. Then, one day, General Wilson received a visitor in his office, Col. Jacksel Broughton. Jack flew as Commander of the Air Force Thunderbirds and was vice commander of the 355th Tactical Fighter Wing at Takhli Royal Thai Air Force Base between September 1966 and June 1967, leading 102 missions against targets in North Vietnam in the F-105. He had just formed the Lift Development Corp, a new company that specialized in Air Cushion Vehicles, and Del Wilson made sure we got to know each other. I cherished my autographed copy of Jack's new book, Thud Ridge.

Col. Broughton

"Thud Ridge" was a contemporary wartime memoir of his tour in Southeast Asia. Broughton was highly critical of the U.S. command structure then directing air operations against North Vietnam. The book resulted from the court martial of Broughton and two of his pilots for allegedly conspiring to violate the rules of engagement regarding U.S. air operations.

Although acquitted of the most serious charges at his court-martial, presided over by then-Colonel Chuck Yeager, Broughton was subsequently transferred to an obscure post in The Pentagon, allegedly as a vendetta because his punishment was so slight. Required by office protocol to work only two or three days a month, he used both his extra time and his bitterness at the Air Force senior bureaucracy and civilian political appointees in the Department of Defense and Department of the Air Force to compose his book *"Thud Ridge"* while he awaited approval of an application to appeal of his conviction to the Air Force Board for Correction of Military Records. After his conviction was overturned and expunged from his record because of "undue command influence," Broughton retired from the Air Force in August 1968 and had the memoir published by J.B. Lippincott.

Col. Broughton signed the copy he gave me in 1969

History Repeats Itself,
By The History Net

Brigadier General William' Billy' Mitchell, a World War I hero and an outspoken advocate of air power, was court-martialed in 1925 for publicly condemning the Navy and War departments for 'almost treasonable' neglect of the Army Air Service. He subsequently resigned from the service and died in 1936. Ten years later, he was rehabilitated in the eyes of the military when, after World War II, Congress voted him a posthumous medal for his 'outstanding pioneer service...in the field of American military aviation.' Today, Mitchell is honored as one of the founding fathers of the U.S. Air Force.

Broughton, too, left the Air Force after his court-martial. During his retirement, he wrote *Thud Ridge* and *Going Downtown*, both books about his experiences as a Thud pilot in Vietnam. While his rehabilitation was not nearly as dramatic as Billy Mitchell's, Broughton was returned to Air Force favor. In 1997, three decades after the *Turkestan* incident, Air Force Chief of Staff General Ronald R. Fogleman directed the Air Force to buy 10,000 copies of *Thud Ridge*, which, along with 12 other books on the Air Force's basic suggested reading list, were provided free of charge to all Air Force officers upon their promotion to captain.

Bridge to Potential Independence

I was emphasizing the Air Cushion Vehicle commercialization assignment. I had developed a relationship with the Canadian Coast Guard in Vancouver, which had three British Hovercraft 25-ton ACVs operating out of the Vancouver International Airport as fire-rescue vehicles.

In the U.S., the market I was targeting was near water airports. However, there were quite a few on the west coast alone, and there with the Canadian Coast Guard, I worked out a demonstration event for as many fire chiefs as we could recruit. Bell's new competitor, Aerojet General, had hired Bell's Wilf Eggington (heading up their ACV division) and Jimmy Sober, who had trained the DoD combat ACV operators for Bell.

Sober, getting wind of my efforts, made Aerojet's corporate jet available to transport those without the travel budget to participate. All in all, we had 15 or 20 fire chief's signed up to go. General Schweizer decided to go with me, and it was fortunate he did.

About this time, ARCO had struck oil on the North Slope of Alaska. It was an obvious opportunity to utilize ACVs for transportation across the environmentally-sensitive tundra (which required low-footprint vehicles less expensive than helicopters and Air Cranes). In addition, Bell was refurbishing all six combat vehicles from Vietnam by then and wanted to sell them to start commercialization. Putting this together, I offered them to develop an ACV operation on the North Slope using the refurbished ACVs. I contacted Bell about my ideas and received management encouragement to proceed.

The Plan

As soon as my idea was in motion, I contacted my second cousin, Cliff Long, in Ft. Worth, to get introduced to his salespeople (retired military) at Century Insurance. He set me up with a retired Air Force Col. in Anchorage who was well connected. He and his wife were great hosts and a big help, taking me to a mountain resort overlooking Anchorage, connecting me with business people, and even having a friend make me the custom gold Indian jewelry (at $32/oz.) I wear it today. Along with Joe Fletcher's Rand Corporation contacts and tutoring, I felt prepared to start exploring in earnest.

My wife, Sondra, the General, and I arrived two days early for the Vancouver demonstration to finalize plans with the Coast Guard and take the downtown commercial ferry ACV operating out of Nanaimo's Ferry Terminal at Departure Bay, Vancouver Island. It was a 25-ton vehicle like the Coast Guard's, built by British Hovercraft. The transit time for its 50-mile ride was less than half the conventional ferry. The Mayor of Vancouver proposed a formal dinner on the demonstration evening.

It turned out that this was General Schweizer's chance to use his French skills to toast the Queen and cap off a great day. Their three ACVs made high-speed simulations of rescue missions, landing reasonably softly on the large cushion of air even after jumping over canals.

The day before, though, the ferry ride to Nanaimo encountered a five to six-foot chop. The rough water required the ACV to cut back from 60-plus knots to 50 or less, a rough enough ride that General Schweizer insisted we return that afternoon on a high-wing, 10-passenger, single-engine airplane that we boarded at the Departure Bay airport. Sondra and I departed the day after, bound for Alaska.

Loading on the single-engine ferry aircraft to Vancouver from Nanaimo

Alaska Kickoff

The atmosphere was boiling with enthusiasm with the ARCO oil discovery a few months earlier, so with the assumption that contracts would be forthcoming, we decided to make a refundable deposit on some nearby acreage for the potential development of the office and hangars site. Having spent time with me in Los Angeles, Joe Fletcher, Rand Corporation's (and the United States) Arctic expert, connected me with a professor at the University of Alaska in Fairbanks.

Next was making contact with the oil industry and government offices in Anchorage. Though more of these were in Anchorage, Fairbanks' proximity to operations on the North Slope was more easily accomplished from there. The results of visits to both cities were encouraging enough that Sondra returned to Palmdale to relieve my parents from caring for the children. Finally, I flew to Buffalo to resign from Bell and make initial plans to purchase their six refurbished ACVs. The ACVs would be partially disassembled for chartered C-130 transport to the North Slope of Alaska.

The North Slope Oil Companies & Transport Development Company

I already had several arctic dealings with major oil companies executives from Bell. I thus returned to them with my new proposal to support exploratory oil operations on the North Slope. ARCO, the first to discover oil there, was using Air Cranes to re-supply the 24-hour/day operations. ARCO focused on the vital rig in an area of Prudhoe Bay with high potential. ARCO aimed to obtain valuable data allowing a competitive edge over many other oil companies that registered their bids due before the September 1969 auction. I had lunch in Dallas with the executive appointed head of their Alaska operations. He agreed that if I could get the ACVs to Seattle, they would transport them along with their other equipment for trials on the North Slope.

Unfortunately, they lost the Air Crane operations and were grounded due to a fatal accident on their rig. Olympic Geophysical (a Houston geological survey company) operated two 25-ton ACVs on the slope. Each ACV had large hydraulic thumpers at the aft ends installed; these would thump the tundra while the instrumentation, strung out in arrays, would measure the acoustic signatures to be analyzed for potential drilling sites. Unfortunately, the thumper operating shocks tore up the ACVs' lightweight structure. Olympic was glad to have ARCO's offer to keep their rig running by hauling supplies, which paid much more than their original contract.

When I tried lining up financing for my new Transport Development Company, one of the ACVs servicing ARCO was returning to the supply depot via a riverbed. They had picked up a hitchhiker (a Canadian geologist) who, upon boarding, stood (unseated or belted in) behind the operators when the tailwind caused the driver to lose control of the ACV and hit the wall of the riverbank. The geologist's neck broke, killing him. The two fatal accidents on ARCO's hands, all within weeks of one other, inhibited launching a new ACV contract for the Transport Development Company.

The few months of frantic North Slope exploration activity were rumored to result in companies' spying on each other's operations, one even using a ground fire to discourage intruding helicopters. For example, Alaska took in so many billions of dollars at the September 1969 auction that they chartered a jet airliner to get the checks cashed throughout the lower 48 states the next day, making much more on interest than the flight cost.

In the summer of 1969, the recently-hired Eggington, tasked to form an Air Cushion Vehicle Department at Aerojet, had me perform a quick market study in Alaska and advise Aerojet on the potential market for ACV production in the Arctic. Taking my advice in my 30-day market study paid by Aerojet, they decided not to enter production based on that market alone.

Final report to Aerojet

1970's Decade

My Hi-Tech Alaska venture developing wheeled air cushion vehicles for the tundra, engineering consulting for the U. S. Navy's 100 knot Surface Effect Ship programs. Teaching renewable energy for the California Energy Commission.

Joe Fletcher
My landlord in Santa Monica canyon 1970-1975

My inspiration for off-road vehicle technology: I met Joe in 1969 when he was at the Rand Corporation in Santa Monica. I was the Bell Aerosytems West Coast marketing representative for electronics and ACV product lines. Joe would show me his 16 mm films taken from his B-29 flights over the arctic as the squadron commander in Fairbanks, Alaska, right after the war. Soon after, I leased his house in Santa Monica from 1970-1975. He made contacts for me in Alaska, including the University of Alaska. He advised me as I attempted to purchase (4) Bell SK-5 Air cushion vehicles re-furbished from the combat trial in the Vietnam war. His oral history interview from the Ohio State University is part of Appendix E.

Dr. M. G. Bekker

Dr. Bekker and his work on the Lunar Rover --Bio in the 1920s Decade

As I developed more definitive proposals, I became convinced that a wheeled air cushion vehicle would make the most sense for both economics and the Arctic tundra's safety. I researched off-road wheel designs and contacted the recognized expert in the field, M. G. Bekker, who lived in Santa Barbara, CA. He and his wife were very hospitable and proud to show my wife and me the technical library of off-road documents he maintained for the U.S. Army in his home. In 1970 I attended a conference in Los Angeles chaired by Bekker on the Lunar Rover. After the conference, he was respectfully critical of the papers and regarded by the industry as having earned the **title of the *"Father of Off-road Mobility."***

<u>Sondra and I Leaving for the SETP 1972 Banquet</u>
Hanna Reitsch Awarded Honorary Fellowship

This was the 25th USAF Anniversary. I attended the 75th in Denver, September 17, 2022.

The TDC Low Cost Wheeled ACV

John Matson, marketing manager for Weber Aircraft (supplier of LLRV/LLTV ejection seats), was looking for a new venture, and we agreed to form a partnership. John was also VP of Engineering for Mitchell Camera (famous for high-speed film cameras and projectors) in Burbank. I convinced him to help me with the wheeled air cushion vehicle idea. We filed for a patent on our new two-psi, large-diameter soft tire and a control system that allowed selective traction control from each corner of the vehicle. Meanwhile, one of their potential customers was Olympic Geophysical (the same company that used thumpers on their ACVs in the North Slope). Their CEO, Charles Robinson, hired me to work at Aerojet on their behalf. We got the patent on the control system but not the tire (due to prior art).

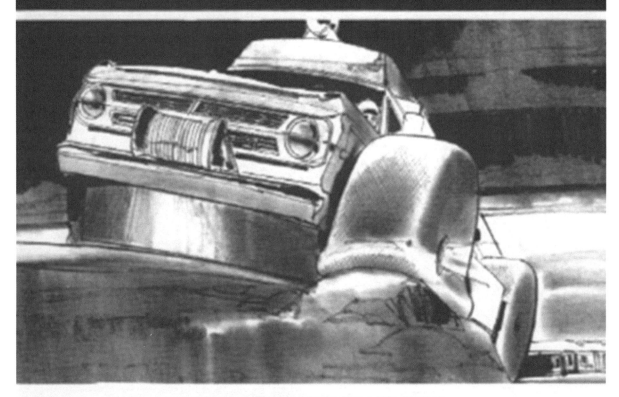

The front side of the Transport Development flyer

TRACAIR

FEATURES OPTIONAL FLOTATION & WATER PROPULSION CAPABILITIES

- STANDARD 4 WHEEL DRIVE AUTOMOTIVE EQUIPMENT + OPTIONS
- POWER OPTIONS OF DIESEL, TURBINE, OR GASOLINE
- AUXILLARY LIFT POWER PLANT (DELETED WITH TURBINE)
- AIR CUSHION SYSTEM VARIABLE TO AUW.
- TRACTIVE CONTROL & PROPULSION WITH TRACON TIRES
- TRACON INTEGRALLY CONTROLLED WITH AIR CUSHION FOR STABLE VEHICLE OPERATION OVER OBSTACLES TO 2' HIGH
- SPEED CONTINGENT ON TERRAIN UP TO 25 MPH
- FOOTPRINT PRESSURE OF .7 PSI FOR AIR CUSHION & LESS THAN 2 PSI FOR TRACON
- PAYLOAD – 2-1/2 TONS

"FOR MANY CARGO HAULING JOBS THE USE OF TRACAIR AS A TOW VEHICLE FOR A TRACON EQUIPPED AIR CUSHION TRAILER WILL FURNISH A HIGHLY MOBILE MEANS OF MOVING HEAVY CARGO OVER SOFT TERRAIN."

- DIMENSIONS: CARGO AREA - 15' x 30' OVERALL - 15'W x 47'L x 9'H
- LIFT POWER OPTION OF DIESEL, TURBINE, OR GASOLINE
- SEPARATE FUEL SUPPLY FOR TRAILER LIFT
- 2 TRACON TIRES FOR STABILITY
- STEEL CONSTRUCTION
- 2' OBSTACLE CLEARANCE
- CONTROLLED FROM TRACAIR TOW VEHICLE
- FOOTPRINT PRESSURES SAME AS TRACAIR
- 12.5 TON PAYLOAD AND 2 TRAILER COUPLING FOR 25 TON PAYLOAD.

TRACAIR TRAILER

FOR PRICE & DELIVERY WRITE OR CALL:
TRANSPORT DEVELOPMENT COMPANY
P. O. Box 919   Santa Monica, Calif. 90406   Tel. (213) 479-0735

The back side of the Transport Development (TDC) flyer

By late 1969 three aerospace companies partnered to form a North Slope oil industry service company, Service City USA; the company had a small refinery to supply fuel for the equipment needed to operate the oil field. John and I met their CEO (a former Lockheed executive living near our Santa Monica Canyon home). His company had committed to TDC a loan of several hundred thousand dollars in financing to develop the prototype of what Matson and I defined as the Wheeled ACV (a four-wheel-drive, one-ton pickup with our low-pressure, large-diameter tires). The funding would become available as soon as they received their $5 million operating loan from the Ford Foundation, expected the following week. The last signature on their loan turned out was from Stephen D. Bechtel, who had the inside track on the looming legal battles for the new Alaska pipeline, the critical pathway to the successful development of Alaska's North Slope oil discoveries. These battles continued for about two years before the U.S. Supreme Court decision permitted the pipeline to proceed. Unfortunately, the day after the disapproval of the loan, the CEO of Service City committed suicide just a few blocks from our house. John and I were shocked.

Dr. Joseph Fletcher
(Biography in 1920s Decade)

Joe frequently visited during the five years we lived in his house (having sold our Palmdale home, we had leased Fletcher's Santa Monica Canyon house in 1970), telling stories of his work at the National Science Foundation (NSF) in Washington, DC and Russia with their Arctic permafrost experts. Three stories stand out. One: He had just come from a conference in La Jolla, CA, at the Scripps Institute, where the attendees were all in despair over research contract cancellations caused by Nixon's major budget cuts. As NSF Polar Programs Chief, his U.S. research ship was on its way to Antarctica staffed with scientists and research gear. It was ordered back to port, but having seen it coming, he worked out a deal to get the U.S. Navy to loan the ship to his Argentine counterparts; they would crew the vessel with the research gear and all the scientists aboard. Unfortunately, they lost the scheduled time for the change, but saved the mission, a fact he was not able to share at the Scripps conference, full of grieving project managers.

A second story involved getting permission from the Soviet Union (in the heat of the Cold War) to take his 14-year-old son for a train starting on the east coast of Siberia and ending at a conference he was attending in Leningrad (In 1991, Leningrad became Saint Petersburg once again). With bags packed on short notice, he and his son were booked on the waiting list for an Air Force transport to Japan. After arriving in Japan, their ship to Siberia went through a typhoon before they could catch the train heading across the Soviet Union. We got a postcard telling of the huge crowd gathered at the China border to watch his son perform card tricks, for most of the Chinese, their first encounter with Americans.

The third story was about when a Soviet Army General (one of their polar experts he worked with on a U.S. tour) and his interpreter stopped by our house in Santa Monica Canyon. We were chatting in our living room when the General stepped over to the fireplace mantle, removed my trophy oosick souvenir from Alaska, approached me, holding it up, and said, "I know what is." Later, Joe confided those were the only words of English spoken by the General on the entire tour. I purchased the trophy in 1969 in Anchorage, AK. Just after the purchase in 1969, I visited the Bell Aerosystems LA office I had left a couple of months earlier, and the secretary to General Schweizer and General Wilson requested that on my next trip, I purchase an oosik she could give her boyfriend (an orthopedic surgeon) on his birthday. I had to call her soon afterward from Anchorage and ask her if she wanted one like mine with ivory carvings at each end or a raw one. Of course, her answer was a raw one; she wanted her boyfriend to guess where it came from. A double bag was necessary, I soon discovered.

LCOL William P. Benedict and LCOL Joseph O. Fletcher in the cockpit of C-47 en route to the North Pole, May 3, 1952. the last few pages of Appendix E tell his story

Four Years Coasting (1970–1974)

Due to layoffs and grim prospects in aerospace, I decided to work for my Santa Monica Canyon neighbor Dick Brewer, who ran WinMar (whose parent company was Safeco Insurance), a commercial real estate development company operating out of Century City. My prime duty was to provide periodic reporting and documentation that all actions were completed to comply with the lease contract provisions. I also did marketing studies for the highest and best use of major real estate investment and condo developments in Palm Desert and Orange County. I earned a California real estate license while working for WinMar, and that experience also assisted in my acquiring a California General Contractor license two years later. A year and a half later, an engineering job opened at the Santa Monica Airport with Lear Siegler Astronics. I worked on fly-by-wire systems for fighter jets, a three-axis sidearm controller for F-16s, and an early space shuttle control system proposal, supporting McDonnell Douglas's bid for the space shuttle, which, at that time, included jet engines for landings (idea eventually eliminated based on the experience with the X-15 and Lifting Bodies). After a year or so, the downsizing layoff came, and soon I was working at North American Aviation at LAX on the B-1A Iron Bird flight control system fixed base simulator (including the landing gear) for another year and a half. The program was canceled just after I was assigned to assist in the flight control system checkout before the B-1A's first flight.

Pentagon Visit

My barber in Santa Monica had a client named Malcolm Curry who, being a single father and moving to take a job in the Pentagon, was looking for a local family to take on his teenage son to room and board. We had him and his son for Sunday dinner to explore their request. Though they chose another opportunity, the experience panned out when later, I needed to make a proposal to DoD for my air cushion vehicle and was able to make my presentation to Malcolm Currie at the Pentagon in his office. Malcolm R. Currie was the Director of Defense Research and Engineering from June 21, 1973, to January 20, 1977.

I worked on the Iron Bird for Rockwell at their Imperial Boulevard LAX (Los Angeles Airport) plant. I was still promoting my Transport Development startup and had a new business contact in the Rockwell corporate office. He eventually arranged for me to meet with Dale D. Myers, who authorized me to travel at company expense to the Pentagon and to the Rockwell Automotive Technical Center outside of Detroit to make my presentations.

My old friend, Norm Foster, had his Space Age Control company and was a close friend of Vice Admiral Forest S. Petersen ("Admiral Pete"), the former X-15 Pilot. Norm had made a couple of hundred unique pilot emergency kits for the X-15 program, and as it turned out that Pete was assigned to the Pentagon to work for Malcolm Currie, Norm gave me one of his new kits to give him when I was waiting to see his boss.

Pete and I had a good relationship during the X-15 program. I had a pleasant conversation with him after he arrived from the first six months of his Admiral Rickover training program in the Idaho Falls reactor facility. Pete competed with another Navy Commander as the Executive Officer (second in command on the nuclear carrier Enterprise). They had been working twelve-hour shifts seven days a week for six months in Idaho Falls and had just arrived to finish the last six months at the Pentagon to end the competition. He enjoyed his alcohol as an X-15 pilot but told me he had not had a drink since he was assigned the Rickover. Nevertheless, Pete won, serving as the Enterprise's Executive Officer from 1964 to 1966. In 1969 Pete was appointed as Commanding Officer, but a fire on the Enterprise flight deck in January 1969 delayed his serving as its Commander Officer till July 1969, during the Vietnam War.

In the summer of 1973, I booked a 747 flight to Atlanta, connecting to Washington National for my meeting with Malcolm Currie. I declared Norm's emergency kit (as it had a .22-caliber gun with ammunition), but the airline personnel could not find it upon arrival in Atlanta. I raised a ruckus and was finally escorted into the 747 cockpit, where I found the kit in the flight engineer's station's cabinet under the console. I got it through the National Airport security and the Pentagon security. I delivered it to Admiral Pete just before my meeting, which, though cordial, never resulted in any DoD interest. Myers was the Associate Administrator for Manned Space Flight at NASA from 1970 to 1974, and from 1974 to 1977, he was Vice President of Rockwell International and President of North American Aircraft Group.

Forrest S. Petersen

SAC (805) 273-3000

ADVANCED SURVIVAL KIT AND SURVIVAL WEAPON

Over 200 items in the kit for the parachute

Obituary:

Vice Admiral Forrest S. Petersen (May 16, 1922–December 8, 1990) was a United States Navy aviator and test pilot. In August 1958, he was assigned duties as Research Pilot in the X-15 Program and served at the Dryden Flight Research Center at Edwards Air Force Base, CA, until January 1962. During that time, he made five free flights in the X-15 and achieved a speed of 3,600 mph (Mach 5.3) and an altitude of about 102,000 feet. He was one of the initial three test pilots with Joe Walker, Bob White, and contractor pilot Scott Crossfield. He was the only active duty Navy pilot to fly the X-15 (John B. McKay, Milton O. Thompson, Scott Crossfield, and Neil Armstrong were former Navy pilots). In July 1962, he was a joint recipient of the Collier Trophy, presented by President John F. Kennedy, and the NASA Distinguished Service Medal, which Vice-President Lyndon B. Johnson presented.

Petersen served as Commanding Officer of Fighter Squadron 154 before being assigned to the office of Director, Division of Naval Reactors, Atomic Energy Commission for Nuclear Power Training. He reported to the aircraft carrier USS Enterprise (CVN-65) in January 1964 and served as Executive Officer until April 1966. He was awarded the Bronze Star for duty during Enterprise's first combat tour in Vietnam. In November 1967, he assumed the USS Bexar (APA-237) command in the Pacific Fleet Amphibious Forces. Following an eight-month deployment with the United States Seventh Fleet Amphibious Forces in the Western Pacific, he was awarded the Navy Commendation Medal with Combat V. He then served as Commanding Officer of the USS Enterprise (July 8, 1969}.

Engineering at Rohr, SES

ADMIRAL ZUMWALT, the U.S. Navy Chief of Naval Operations, was campaigning for a 100-Knot NAVY. The Air Cushion Vehicle industry, started in the 1950s by Sir Christopher Cockrell, had always envisioned both the fully amphibious type of watercraft (operating in water and land) and the surface effect ship type where its hull could rise mostly out of the water. Still, the supporting air cushion would be mostly contained by solid sidewalls, flexible bow, and stern inflatable bags. This design resulted in much heavier payloads at high speeds to be carried by a powerful water jet propulsion system. Since my last two aerospace adventures proved tenuously short, I consulted with Rohr Industries on the surface effect ship program.

For several months, I performed design study tasks on their 3,000-ton destroyer class prototype. Jimmy Sober, the former Bell Aerosytems trainer for the six Army and Navy ACVs in Vietnam river combat, managed their 100-ton test craft Rohr support facility. It was located at the Navy's Flight Test Center at Lexington Park, MD (where the Patuxent River enters the Chesapeake Bay). Sober needed more engineering support to add to his staff of technicians and mechanics. My consulting contract changed to work for Jimmy, including producing the periodic film reports for the Pentagon on the test craft program.

The most rewarding engineering challenge I worked on was selecting the material and fabrication method to install a custom geometric shape to the bottom of the sidewalls of the SES 100A. The shape was determined by experts running scale-model tests in the Navy's hydrodynamic testing facility (water tunnels) near Washington, DC. The goal was to improve efficiency by reducing drag caused by the turbulence of the air/water mixture at high speeds at the bottom of the sidewall.

My research started at Hercules Powder, the manufacturer of the material. I chose an ultra-high molecular density polyethylene material for the first successful replacement of human hip ball/socket joints. This material was also used as bearings for the main propeller shafts on nuclear submarines due to the low friction, corrosion resistance, and, even more importantly, the low noise signature crucial in submarine operations. Hercules allowed me to survey various fabricators, selecting extrusions suitable for machining and attaching to the bottom of the sidewalls on the SES 100-ton, 100-knot ship. It turned out to be a fabricator in York, PA, the town selected to settle in by Johaann Jacob Ottinger (my oldest U.S. ancestor, 1716–1781)

The 3,000-ton Destroyer Class Rohr Industries concept

The SES 100-ton test craft

The film reports for the Pentagon required reviewing the given periods' technical achievements with Rohr and Navy engineers. In addition, the film reports involved organizing the 16-mm film shooting requirements for the Navy cameraman, script writing, graphics production, editing, and approval by both Sober and the Navy Commanders operating the SES before delivery to the Pentagon. The work led to other film productions, such as an educational film on the history of air cushion vehicles and surface effect ships shown at Navy orientation classes, produced for Jimmy Sober and the Navy's Flight Test Center at Pax River.

Another assignment was for Admiral George Halverson, a promotional film for the 3,000-ton SES designed in Chula Vista. Through Sober, I obtained the entire British Petroleum library of films documenting the early days of Sir Christopher Cockerell's ACV and SES inventions and developments. I also hired a local San Diego consultant, Wil Berg, who had animation and filmmaking experience. We purchased some essential animation tools, having the Rohr artist generate artwork. Between the BP film library and our stock footage, we produced the eleven-minute "*New Way to Go*" film for the Navy, delivering it to the Amphibious Hovercraft Base on Camp Pendleton in the early 1980s. Both films were produced in 1977 from the single garage at Esther's (my second wife's) condo where I lived. Sober wanted his orientation film to be kept under wraps so he could surprise Halverson with a complete work; we had quite a time removing Sober's materials from sight (particularly the animation) during the Admiral's visits to the garage.

My frequent trips required me to go through Dulles International Airport. There I would stop by a road stand selling freshly-picked corn, buy a full sack, and put it in the overhead on the flight home to barbeque that night. On one trip, I was in a rented Ford Pinto on a two-lane road a few miles from Lexington Park, just after the bars had closed. Unfortunately, the oncoming car lights just ahead were in my lane; I suddenly braked hard, and to this day, I have no idea which side the oncoming car passed me, as I had been spinning out of control. The following day I switched to a full-size car.

A similar incident occurred in 1985 on the I-405 Interstate: My daily round-trip commute was 186 miles for five years, and thus I had developed bad driving habits. On my way home from work in Torrance, at over 80 MPH, I suddenly braked hard, seeing brake lights and stopping traffic a ¼ mile ahead. The front-wheel-drive

Accord spun through all six empty southbound lanes and faced North next to the freeway fence with a dead engine. I checked my pants immediately, seeing the large puddle of water below the empty passenger seat, but realized I was looking at the results of the "g" forces sucking out the condensate from the air conditioning vent in front of the passenger seat. Within days I had a new Lincoln Town Car and toned down my driving habits.

My first marriage ended in the mid-70s after many attempts to reconcile, and soon after, my second wife, Esther, 35 years later, passed away in 2010 after her stroke in 1998. After that, however, we, for eleven of the twelve years, traveled happily with the wheelchair.

Esther Meets Hannah Reitsch

Biography in the 1920s Decade

In 1975, we attended the Society of Experimental Test Pilots Symposium at the Beverly Hilton in Los Angeles. The Saturday morning session "Oldies But Goodies" featured the old timers and Hannah Reitsch as a keynote speaker. Her record as a German Nazi Test Pilot in WWII was recognized by the SETP in 1972 when she was inducted as an honorary member; Sondra and I were there at that banquet, too. Esther had just completed a few years as the affirmative action officer at the Grossmont Community College in San Diego. She was active in getting the unfair rejection of women candidates for hire as faculty changed. She was fortunate to hear Hannah's story of her outstanding record of service in the male-dominated Nazi military. It was a privilege for all to hear her discuss dogfighting tactics (with old British fighter pilots in the audience) that neither side was aware of during the air war.

California Energy Commission (1977–1978)

After working for Rohr, I applied for a position at the California Energy Commission in Sacramento just before we left a mountain vacation home in the Sierra's in September 1978 (a month after Elvis Presley died). It was expanding its energy conservation and renewable energy program staff. I was assigned to work for Warren D. Hinchee, who had just left his short tenure as manager of the Burbank Municipal Utility.

Warren's unique background included serving in the Army in WWII and working for the CIA during the Bay of Pigs invasion of Cuba (where he was finally rescued from the jungle two weeks after being shot in the back). His management skill was eclectic, a trait I wish I'd learned earlier. His responsibility focused on energy conservation, handled by a crew of experienced industrial engineers, who turned into energy auditors who would visit companies and recommend positive corrective actions to save energy. Warren told us that just before his Burbank job, he worked as the municipal utility manager for Dennis Kucinich, Cleveland's very young mayor. In the heat of the dispute over the take-over attempt of the city-owned municipal utility, the going got rough. The young mayor was reviewing papers in his ground-floor apartment one evening and had leaned over to pick up more documents from the floor from his rocking chair. At that moment, a rifle shot was fired through the window, targeted at the spot where his head had been.

The educational workshops organized and held for municipal utilities, condo associations, and local governments included energy conversation and renewables such as solar. My recent movie production work for the Navy led Warren to accept my proposal to produce an energy conservation film for the energy commission. He and I were both new to the state government protocols, but we charged ahead, working with the State Water Department's movie production unit. I hired Wil Berg from San Diego, who helped at Rohr, to help me with the animation and production of the 13-minute film *The Phantom City.* It was a big hit with the commissioners; one of them, Dr. Ronald D. Doctor, had me accompany him on a trip to Washington, DC, to lobby for federal grants.

Esther and I bought a new house in Loomis (between Auburn and Folsom) with a modest dock in the backyard facing a small lake, handy for swimming or fishing after work. My youngest son, Bob, was finishing high school and starting community collegein the area and was able to stay in our new house for several months until we sold it after moving to Long Beach in late 1978.

While in Washington, DC, with Dr. Doctor, Esther and I found tee shirts we bought from a sidewalk vendor. Esther's said, *"When God Created Man, She Really Didn't Mean it!"* Mine said, *"It takes a Mighty Secure Man to Love a Liberated Woman."* After arriving back in our new house in Loomis (east of Sacramento), we wore those shirts the next day and drove to Oregon to rendezvous with Esther's ex-son-in-law and pick up her grandchildren, Denise and Eric, at Rogue River Park (near Grants Pass). Even though we were pretty late, arriving back in Loomis at 2 a.m., there were 20 or 30 of Esther's family members patiently waiting to surprise us with a housewarming party. The party resumed about 10 a.m. the next day, with both their owners' tee shirts proudly worn.

The Department of Energy assigned a psychologist from Stanford to work with me in designing surveys/ educational material and conducting seminars to offer design options for Friars Village, the 430-unit, an all-electric condominium project in San Diego. Esther's condo was part of Friars Village. We used the State's movie production unit to film the HOA meetings, which included a scale model of a centralized solar system serving the entire project. I recruited my old partner, John Matson, from the Transport Development Co. to build the model. As this project proceeded, the new Solar Department in the California Energy Commission was maturing, and the bills for movie production were starting to land in their lap. Unknown to me, Warren Hinchee was about to leave for an outside job, apparently unmotivated to get involved in the budget battles. This situation dictated that I resign and not try to earn a position in the new Solar Department, which had conflicts with Warren.

1980's Decade

My stories for this Decade are diagnostic imaging, motion analysis for advanced high-speed large centrifuges for the Department of Energy, and my small business applying these technologies for industry.

Garrett AiResearch (1979–1987)

We moved to the Los Angeles area after I resigned from the Energy Commission. Esther partnered with her sister, Marge Holmes, who lived in Belmont Shore (south Long Beach), to start a lingerie and prosthesis specialty shop for breast cancer survivors. Marge was a survivor and envisioned the need for such a new business and collaboration with the CEO of Mattel, Ruth Handler, who had started a business (Ruthton Corp) after her bout with the disease. Their retail store, ideally located near Leisure World in Seal Beach, was called LaVail (Vail being the name of their maternal grandmother).

I secured a position at the AiResearch Corporation in Torrance in the Gas Centrifuge department, developing much-improved uranium enrichment centrifuges for nuclear power plants. The project required a "Q" clearance from the Department of Energy (DoE). Fortunately, I had a temporary assignment for several months in Phoenix while awaiting security clearance. Once cleared to work on the project, I soon found a good niche in developing diagnostic tools using special optical and video inspection equipment and producing training videos for the department. The company was part of several large corporations working on developing gas centrifuges, and both common and proprietary technology was involved. The development program had been going on for over a decade. I worked there from 1979 to its termination in 1987, just as mass production was to start.

Soon after moving to Long Beach from Loomis, we bought a 1979 diesel Oldsmobile four-door sedan. GM modified a standard V-8 gasoline engine and put a 28-gallon main diesel tank in the large trunk, playing on high gas prices to sell these models. I took the chance, even adding two auxiliary 25-gallon diesel tanks for a total of 77 gallons. Unfortunately, our 26-foot fifth-wheel trailer storage lot rental fee was excessive in the Los Angeles area, so we found a KOA campground in Baja, CA, overlooking the beach just north of Rosarito Beach, which cut our rent there (with full hook-ups) to less than half the U.S. storage costs. We used the trip there for several years to fill the Oldsmobile with fuel at 18 cents/gallon amidst long lines of Americans filling up in Tijuana. I had to buy truck filters, though, as the fuel was very dirty and would easily stall the engine. We made two round trips from LA/San Diego to Phoenix on one fuel load. By the time I had 95,000 miles on the Oldsmobile, the head gasket had shown signs of leaking, and I sold it, acquiring the Honda Accord for my long commute from Vista to Torrance.

Trip to Europe

I had scheduled our trip to arrive in time for the air show in Hanover, Germany (which alternates every two years between Germany and London). At the air show, I took my remarkable photograph on Kodachrome 35-mm film.

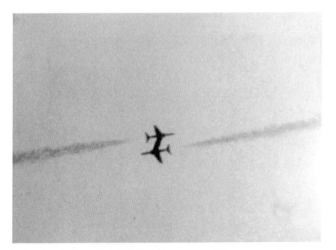

1980 air show, RAF Red Arrows, Hanover, Germany

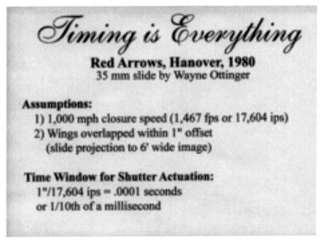

One-tenth of a millisecond

Expanding Motion Analysis

My most challenging task at AiResearch was to develop a high-speed imaging system to accurately determine the location of failure in newly-designed parts rotating in the spin tanks. The four-foot diameter steel vacuum tanks were used to test the components until they failed at high speed, exploding. Unfortunately, no film-based imaging system was capable of documenting the failure, the visual evidence needed to determine design fixes. With about 50 engineering jobs on the line, I was fortunate to find a solution pretty quickly. With the recent development of 2,000-frame-per-second video tape recorders (Kodak's Spin Physics division in San Diego), the copper vapor laser (strobe light), several hundred thousand dollars, and a lot of innovative work, we succeeded in getting the first pictures of the explosions in about three months. Steve Picarello (the lab technician assigned to me) was so good that the laser company Plasma Kinetics hired him a year later. Steve now owns his laser repair business, Expert Laser Services, LLC, with domestic and international customers.

Near the end of the gas centrifuge program, my TV crew and I spent a few days in Bowling Green, KY. We videotaped the old Firestone Tire plant, which, having been empty for years, AiResearch considered a potential production site for gas centrifuges needed by the DoE. After the Or Firestorm film was edited and presented, soon after, DoE canceled the entire program. The demand for enriched fuel dropped significantly after Three-Mile Island and Chernobyl. With the old, inefficient gas diffusion plants wearing out, a company in southeastern New Mexico is finally firing up new gas centrifuge production. The 40 or 50 billion government investment dollars spent in the 1960s on gas centrifuges will not all go to waste.

The HR department at AiResearch sent me on to work for the United Way in Long Beach for six months as a loaned executive after the DoE project was shut down. Before leaving AiRsearch, I spent another year in manufacturing engineering until a RIF (reduction in force) caught my seniority, and I was laid off.

<u>Spin tank high-speed video/laser strobe failure diagnostic system, 1985</u>

Close-up of two-inch-thick steel lid with Lexan imaging ports

1984 Olympics

Dave Drees, who ran the video production of my group, took a video of the torch relay that ran in front of our plant on 190th Street in Torrance. The principal vendor for our video lab in Los Angeles was Mac Brainerd, a senior engineer that had been part of the WWII MIT team perfecting radar. He picked up Esther and me in his Rolls Royce for the Concert gala at the Hollywood Bowl.

During my tenure at the gas centrifuge project, I became close friends with Charles Stenning (of Stenning Instruments, one of our vendors), whose expertise in diagnostic imaging tools was superb. He was the son of the British head of the Bank of England in South America, and his mother was a native Indian from Peru. Charles immigrated to the U.S. in the 1950s and was self-taught in optics with excellent industry ties in Japan and Germany. He supplied me with borescopes, fiberscopes, lenses, and various video and microscopic tools. Charles also had a unique background starting WWII as a young infantryman in the North African campaign. Fortunately, he pulled out of combat duty at age 19 to replace a British intelligent agent killed in South America. In addition, Charles had learned six Indian dialects during his early life there in South America.

SynerVision (1988–1990)

I received a Video Probe for the 16-minute Welch Allyn sales videotape I produced. Using these systems and my experience in diagnostic imaging, video production, and graphics, I formed a general partnership with Erv Metzgar (retired president of Grossmont College). I trademarked Synervision providing imaging services to clients such as Kellogg's, McDonnell-Douglas, and the San Onofre and Palos Verde Nuclear power plants. In addition, I hired graphics production personnel, acquired the color thermal printer for viewgraphs, and supplied dedicated production services for conferences held at Edwards AFB NASA DFRC and the NASA Ames center in Mountain View, CA.

The U.S. Army Proving Ground in Yuma, AZ, requested bids for a mobile high-speed video/laser imaging system to test ordnance. The open competition advantage I thought SynerVision had as a small business and the spin tank experience at AiResearch was worth the three-month bid expense. The AiResearch system was the first industrial use of the laser strobe, with the high-speed video decks added to my confidence for its use in SynerVision. Unfortunately, after the bids were accepted, the Army's funding was lost, and thus so were three months of work and expense. Thankfully, I had sold my video probe system to the San Onofre Power Plant by then. They and the Palo Verde Power Plant purchased their systems after their satisfaction with the results of my services.

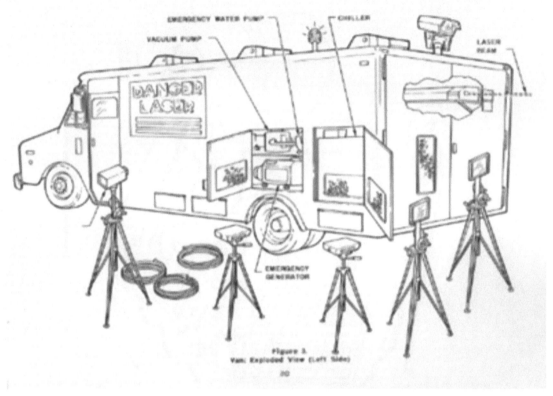

Proposed mobile strobe light facility for the U.S. Army Yuma Proving Ground

The Remarkable Passive Solar Home (1980–1988)

It was 86 miles from my Torrance job, but I knew I could handle the commute for a couple of years (stretched to seven years, it turned out). We hired a reputable architect experienced in passive solar designs and moved in late 1980. The 2,400-sq.-ft. all-electric house had no propane, and the total energy bill averaged $35/month. I instrumented the house with 14 thermocouples, taking hourly readings for 14 months.

The few weeks' data analysis demonstrated the performance of the Trombe walls, which retained many times the heat compared to a test cell complex of similar walls in the area run by San Diego Gas & Electric and Southern California Gas. We had one of only two time-of-day electric meters in San Diego County to provide low electricity rates off-peak. The architect viewed the efficiency achieved as a threat to the utility. All we got from the project besides a good living environment was good publicity from Sunset Magazine, the local TV stations, and newspapers.

Passive solar home's thermal flow:

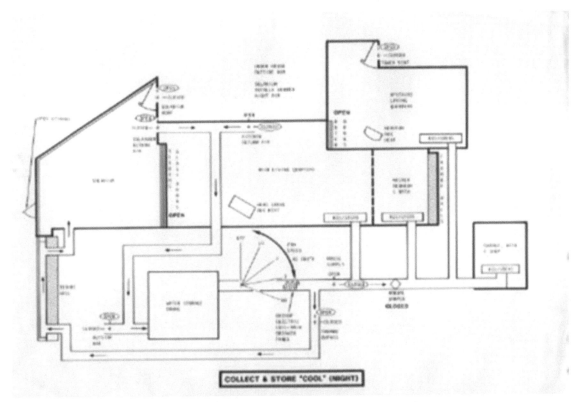

Passive solar home, Oceanside/Vista, CA, early 80's

Thermosiphon Collector downslope from spa

Passive solar house solarium/spa

Esther and spring ground cover at the solar house, the early 1980s

June 8, 1989 X-15 Technical Symposium Pilot's Panel

Eight of the 12 X-15 Pilots were recorded for almost two hours on the 30th anniversary of the first X-15 flight in 1959. Esther and I attended and met Jim and Betty Love. The YouTube link: https://youtu.be/pjZ0 at the 1:12 to 1:20 time in the 1:54 total is the story told by Maj. General Joe Engle (USAF Ret), screenshot below. He is the only X-15 Pilot still living today.

Screenshot of Joe Engle, 1988 Symposium

1990's Decade

I founded nonprofit education for teaching NASA flight research programs, worked on the OV-10 Bronco aircraft, and recalled Whitey Whiteside.

Garrett Engines (1990–1993)

In 1990, the Garrett Corporation Engine Division (formerly Garrett AiResearch) in Phoenix brought me back to work on turboprop engine control systems for the Rockwell OV-10 observation plane.

They went through three layoffs in 1992, finally catching me on the last one, but by that time, I was at least vested for a small retirement and medical insurance. During that time, the OV-10s, based out of the Marine base at Camp Pendleton, were used on observation missions. For example, to get four to five special operations paratroopers behind enemy lines, the OV-10 would climb vertically, open the cargo door, and then the men would fall out of the back end aft of the cockpit. Unfortunately, during my visits to Camp Pendleton, two OV-10s were lost in the Desert Storm combat.

OV-10 Bronco

Nonprofit Education/PAT Projects (1994–2002)

The DoD had issued a competitive funding announcement to support the Legacy Project, a history of the technologies and the Cold War. I visited chief historians Richard P. Hallion and Roger Launius, respectively, at the Air Force and the Smithsonian). I received encouragement to document the stories of the Lifting Bodies, which developed in the 1960s and early 1970s.

In June 1993, we moved to Lancaster, CA, staying initially with Esther's daughter, Rhoda. Marta Meyers hosted Esther and me at the NASA DFRC facility for an early morning Shuttle landing. The landing was scrubbed for bad weather, but Marta had convinced Esther and me to explore the Tehachapi community to get settled. Marta and her husband, Bob Meyers, were stunt pilots and NASA engineers. Marta and her husband were the two of the flight engineers to fly in the back seat of the SR-71.

We took Marta's advice and found a rental in a gated mountain community about 15 miles west of Tehachapi. After having the furniture delivered there, we started to organize our nonprofit, the Preservation of Aerospace Technology, or PAT Projects.

The board of directors was formed by old NASA and Air Force acquaintances and with funding approved by the Legacy Project. The surprise was that the project and funding were assigned to an existing government contractor who immediately claimed two-thirds of the funding, deciding to use PAT Projects as a subcontractor. Unfortunately, Rhoda was working in the subcontractor's contracts department and had to withdraw her involvement because she was my stepdaughter.

Nevertheless, despite a dispute with NASA over some petty interpretations of history, we did get the job done. Soon after the first contract, NASA solicited new proposals for partnerships with schools and nonprofits for internet-based math and science education related to current aerospace projects. I contacted the Palmdale and Antelope Valley High School districts, and we built a team to respond with a proposal we named SHAPE (Sharing Aeronautic Projects Electronically). NASA awarded the funding, selecting the high school districts as fiduciary partners to administer the grant. Working with K-12 teachers and administrators, PAT Projects had test pilots and engineers prepare suitable material, such as the last project for SHAPE, a lesson based on the Space Shuttle 747 Carrier Aircraft (SCA).

For the life of the Space Shuttle program, the SCA transported the Shuttle on its top from landing sites and back to the Kennedy Space Center in Florida to be prepared for the next launch. Gordon Fullerton, an early

Space Shuttle Command Pilot, had left the astronaut corps and joined the NASA Dryden Flight Research Center at Edwards as a test pilot. He also flew the 747 SCA and provided taped lectures for the students. Just after takeoff with the Space Shuttle on top of the 747, an engine caught fire on the right wing. Fullerton had to make an immediate go-around (that is, flying low through a 360-degree turn back to the takeoff spot where the fire trucks were located. He was able to grease the heavy SCA (which far exceeded the landing weight limits) onto the runway and allow the fire crews to extinguish the engine fire quickly, thus saving both aircraft. One of PAT's board members, Lt. Col. Steve Stowe (USAF Ret.), authored the interactive lesson for the SCA mission, adapting an emergency into the lesson plan based on Fullerton's experience.

The SHAPE program delivered substantial work Stowe had completed, including the flight operations planning and execution to prevent the Space Shuttle tiles from encountering rain during the ferry trip across the country. The SCA could not fly over the weather. However, the ceiling of the SCA with Shuttle was about 20,000 feet. So a smaller jet transport would fly about 30 minutes ahead to allow the SCA to divert around the rain, frequently headed for a pre-determined airport equipped to handle the unique needs of the Space Shuttle on its transit stopover.)

NASA Shuttle Carrier Aircraft taking off with the Space Shuttle from Edwards AFB

Rhoda left her job at the Edwards contractor and worked for me for a few months at PAT in our offices at the Society of Experimental Test Pilots Building in Lancaster. We worked out procedures with NASA's HR director and chief accountant to establish PAT as a qualified IPA (Intergovernmental Personnel Act). A long-standing act passed by Congress decades before the 1990s allows local, state, and federal governments and specific nonprofits to exchange personnel for prescribed periods to facilitate cross-training. NASA could fund IPAs to hire former NASA executives/engineers on our payroll. Meanwhile, NASA had given us a grant to write a NASA Monograph on the LLRV program.

After the first year of certification, the IPA payroll had grown to over a million dollars per year, far exceeding the initial expectations. In addition, Rhoda had started working for NASA contracts, again having to recuse herself from our work for NASA. To help meet our increasing overhead shortfalls, the HR contact who started our IPA work was able to secure an assignment for us to hire outside consultants to advise NASA DFRC on strategic planning. The arrangement went well until she alerted us that the NASA legal counsel was disputing the legality of the travel funds paid to one of the first retired IPA executives. Before their travel, the center director approved all the questioned travel orders in writing, so we were shocked at the bureaucratic injustice.

10/8/99 LLRV 35th Anniversary Conference SETP HQ (LR) Warren North NASA JSC, Don Mallick, LLRV Test Pilot, Gene Matranga, LLRV Project Mgr, Bruce Peterson Helicopter Chase, Cal Jarvis LLRV Controls

For several years there had never been an audit requested, so though funding was cut off from NASA, PAT continued to work for NASA free for several months while supplying materials for the audit. Unfortunately, no resolution was ever achieved, so the entire board resigned, and whatever assets were available were used to close the nonprofit's operations. The NASA IPA problem that started with PAT soon spread to the NASA Ames Research Center, which had also begun to use PAT.

Before the IPA fiasco, the NASA education program director was replaced, and unfortunately, most of the teachers' lessons were never published. He was a brilliant software/aerodynamics engineer, and it was a difficult transition for all concerned. Despite the disappointments, PAT did complete many assigned tasks successfully for NASA.

In 2002, PAT partnered with the SETP to raise funding for a new Antelope Valley complex and the centennial celebration of the 100th anniversary of the Wright Brothers' first flight. PAT hired marketing consultant Burnet Brown to assist in developing a campaign to solicit support from aerospace corporations and philanthropists nationwide. As a result, I appeared before both city councils in Lancaster and Palmdale to locate a new SETP headquarters building, including aerospace education facilities. In addition, a new museum along the dividing line between the two cities called "Avenue M." Bud Ream, a retired local school superintendent and PAT board member, took me in his helicopter to acquire aerial photos for the proposed site to use in the promotional literature. Brown's solicitation letters were prepared for SETP stationery and were signed by the president of the SETP at the time, Jimmy Doolittle III (Ret Col. USAF). The night I was with Jimmy getting his signatures, he deferred to Paula Smith, the SETP Executive Director, to sign the letter to Disney's Chairman, Michael Eisner. Jimmy had just had a difficult time with Eisner over the negative implications of the script for their movie Pearl Harbor, which, he felt, reflected negatively upon his grandfather Jimmy Doolittle's personality. Jimmy never looked at the movie's final release but said his cousin reported that some, but not all, changes were made in their favor. They appealed then to Jack Valenti, President of the Motion Picture Association of America, who said the change requests were futile). Paula Smith, also a PAT board member, was extremely helpful throughout the several years we shared the SETP offices.

The Joe Walker School Project

PAT provided the Joe Walker Middle School in Lancaster, CA, with a one-year series of lectures named Technology Lab 2000. Two LLRV crew chiefs (Ray White and Bill Wilson) and Dave Stoddard from the Rocket Shop worked with the lab class to complete a seven-team student competition to build a winning scale model of the LLRV. Joe Walker's widow, Grace, had donated the scale model Bell Aerosystems had given Joe to the Lancaster school, but it was stolen from a display case there. The Class of 2001 decided to replace the stolen model, and I promised the school a tile mural for the school library of the project if they could come up with a good quote for the mural. Their selection, "Learn Math and Science, One Small Step for Schools Across the Nation, One Giant Leap for Our Future," earned the mural donation.

The mural featured photos and signatures of both Joe Walker and Neil Armstrong and the centerpiece of the Norman Rockwell painting of Apollo 11 on the moon, with the background showing pictures of the students in the lab and the PAT tutors. With permission from the Rockwell family, all students are given canvas copies of the mural and the certificate.

5' X 9' tile mural (made by Greg Ottinger, 2000) donated to the Joe Walker Middle School library

The Joe Walker Middle School Technology Lab 2000, Class of 2001 Lunar Landing Research Vehicle Model Project

April 26, 2001 — PAT Projects, Inc., (PAT)

Why America Needed the Lunar Landing Research Vehicle

There are no runways or smooth landing areas on the moon.

Among all of the engineering problems encountered during the Apollo Period in the 1960s and 1970s, the lunar landing posed one of the greatest technical challenges. How would the astronauts touch down safely on the Moon without any of the familiar piloting guideposts available on Earth—without runways, visual reference points, or navigational aids? Eventually, the answer became clear: the landing must be a vertical one.

But having drawn this conclusion, Apollo planners faced a second, more daunting hurdle. How could designers determine the required flight control characteristics for a vertical lunar landing and how would the astronauts practice vertical landings on Earth? Because the gravity on the Moon is only one-sixth that of the Earth, the weight of a vehicle flying above the lunar terrain would be only one-sixth of its value on Earth. Helicopters—although vertically oriented—could not fly on the earth as if they were in the Moon's gravitational field.

Moreover, because the Moon has no atmosphere, a lunar landing trainer needed to handle as if it were flying in a vacuum. Therefore, before the Lunar Module design was even begun, the concept of a free-flight lunar landing simulator to fly on the earth was born in 1962. This concept evolved into a flying machine unlike any ever fabricated, the Lunar Landing Research Vehicle, or LLRV (illustrated in the photograph, upper right).

Utilizing an ingenious combination of attitude rockets and a jet engine mounted on a tubular frame, project engineers at NASA and Bell Aircraft supervised the construction of the LLRV, and then organized a series of research flights at NASA's Flight Research Center in Edwards, California. These simulated missions accomplished the crucial objectives of producing

critical design information for the real Lunar Module and providing Neil Armstrong and his successors with the subtleties of maneuvering the Lunar Module the last few hundred feet from the lunar surface. In so doing, the LLRV and its successor trainer, the Lunar Landing Training Vehicle or LLTV, proved to be crucial in the overall success of Apollo.

The Joe Walker Middle School LLRV Project

PAT teams aerospace professionals with teachers, who working together develop math and science lessons using hands-on learning projects based on their first-hand experiences. Together, they emphasize decision-making, to inspire students toward careers that integrate math and science and communications.

The late Joe Walker, NASA Flight Research Center Chief Test Pilot and first to fly the LLRV, had been given a model of the LLRV by NASA in the 1960s. Neil Armstrong had served under Joe Walker as a research test pilot prior to becoming an astronaut. Joe's widow, Grace Walker, in the 1970s donated the LLRV model to the school. Several years ago the model was stolen, so the current Technology Lab 2000, class of 2001, decided they should build a replacement. The teacher, Matt Anderson, contacted NASA who referred him to PAT.

Together, the PAT team including Matt Anderson and Dave Stoddard, former Rocket Shop supervisor at NASA, developed a learning experience project involving student participation in a contest to design and construct a replacement for the original model. The winning model, shown in the accompanying montage, was a result of the class working in teams using photos and drawings to build the seven models. The team that built the winning model included: Matt Cline, Eric Haviland, Zach Rosen, and Brandon Wood. The judges for the model competition were the former LLRV crew chiefs (Ray White and Bill Wilson) and Dave Stoddard. The montage conceptual design was from Wayne Ottinger, President of PAT (and former LLRV Project Engineer for NASA), and the computer graphics work was done by Karl Jackel and Billy Murrell.

<u>Canvas certificate given to each student with the canvas mural</u>

Before closing PAT, I was contacted by Roger Launius, NASA's Chief Historian, who was visiting NASA DFRC then. At our first meeting in 2003, Roger suggested that whatever funding I could raise to start a nonprofit would be better used to increase his history office budget at NASA. Initially, I had credited his comment to his inexperience as a new Ph.D., but after he came to my home to meet years later, I was satisfied PAT had earned his respect. Our group included John McTigue (a noted DFRC engineering manager and PAT's Chairman of the Board, Bruce Peterson (NASA Lifting Body Test Pilot), and Bill Dana (X-15 Test Pilot), whose wife, Judy, was on the PAT Board).

Nine-sq.-ft. LLRV 30<u>th</u> Anniversary Mural, 1999, I'm on the right in the top crew photo and 2<u>nd</u> from right in the bottom photo. I was 31.

Walter (Whitey) Whiteside

Graveside Revelation, First Impressions Need Restraint.

I briefly encountered the retired FRC assistant to the Center Director, Paul Bikle, Walter (Whitey) Whiteside, in the late 1990s. Unfortunately, I was late to his graveside ceremony in Palmdale, CA. He was the Col. in Maintenance at Edwards AFB when Paul Bikle was the Chief Engineer until he replaced Walt Williams as Director of NASA FRC in September 1959 (7 months before I joined NAS). Walt was later appointed to direct Project Mercury. Paul brought Whitey to FRC, who had been with him since WWII as they worked together to extend the range of the B-29's on Okinawa as the war closed.

Whitey was a motorcycle enthusiast and added their use on the flight line for pilots to get to service areas remote from the hangars. The early flights of the Parasev, the Rogallo Wing tri-cycle, and lifting bodies were Bikle assignments he relished in. Here are several photos of this step-by-step flight research culminating in B-52 launches of supersonic lifting bodies.

The Parasev Towed by a Stearman, Whitey on the left

Rogallo Wing, Evolution from Early Research Vehicles to Hang Gliders and Highly Maneuverable
Parachutes and Space Reentry Vehicles.

Rogallo Wing Evolution

NASA's Armstrong Flight
Research Center
January 22, 2015 · ⊙

#tbt January 29, 1963

Walter "Whitey" Whiteside went to Long Beach
to pick up a modified Pontiac Catalina tow car
for the M2-F1 lifting body research vehicle. The
Pontiac was capable of towing the 1,000-pound
M2-F1 at speeds of up to 110 mph on the dry
lakebed at Edwards. The car, which may have
been the only government-owned hot rod
convertible, was equipped with a specially
modified gearbox to improve transmission gear
ratios. It also had roll bars, an observer seat,
racing suspension, and radio communications

Whitey's Proud Achievement

In the late 1990s, I left my *PAT* office in the Lancaster, CA Society of Experimental Test Pilots (SETP) corporate headquarters and arrived late for Whitey's burial ceremony. I parked downhill and walked up the hill to join the group of test pilots and others as Fitz Fulton was speaking about Whitey about his working with him supporting his heavy bomber flight testing. I had only been there a few minutes, and many motorcycles arrived, disbarring very close to the gravesite. Their group asked for a chance to speak. It changed *the whole mood* as they told about Whitey's mentorship and counseling as he led their club in AA recovery programs with years of dedication to anyone needing help. The respect of their sharing their stories about Whitey offset any negative feeling of their noisy late arrival. Whitey reportedly had driven the Pontiac to 160 mph with no tow.

Betty Love, a close friend of my wife, Esther, and I volunteered for many years in the history office at the NASA Flight Research Center. She interviewed Whitey in his last few years, and I learned he was dropped off as a teenager at the San Diego North Island Air Station in the early 1930s.

Historic California Posts, Camps Stations and Airfields
Naval Air Station, North Island
(North Island Aviation Camp; Rockwell Field; Naval Air Station, San Diego)

NAS North Island circa 1930s. Note the presence of the USS Langley, the US Navy's first aircraft carrier docked at the island.

NAS North Island circa 1930s. Note the presence of the USS Langley, the US Navy's first aircraft carrier docked at the island.

His family struggled in the Fresno, California region in the early depression, thinking he would be better off joining the US Army Air Corps even though he was not old enough. He got in quite a start, winding up in WWII as an officer assigned to Paul Bikle, a civilian engineer from Dayton.

2007 Constellation Program to Return to the Moon

After seeing a local Phoenix TV interview about the details of an avionics contract Honeywell had just been awarded (the Crew Exploration Vehicle), I contacted William G. Gregory at the Honeywell Aerospace facility in Glendale, AZ. After several meetings there, I proposed we explore putting on a conference, eventually named "Go For Lunar Landing." I contacted VPs at Arizona State University (ASU) in Tempe and the University of Arizona (UA) in Tucson. Gregory and I met several times with ASU VP Rick Shangraw, who agreed to take a lead role in developing the conference.

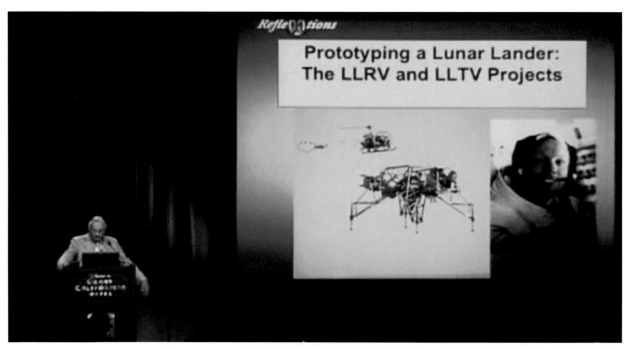

September 29, 2007

In July 2007, Neil Armstrong contacted me to help him prepare for his talk on the LLRV and LLTV in the opening session. I provided him with photos, movie clips, and technical data he needed for his 32-minute presentation. The former NASA FRC contracts manager Lloyd Walsh joined me that Saturday morning.

Joining him at his table after his talk, we had lunch, and I proposed to him to organize a future conference to discuss the challenges for the new Constellation lunar lander named Altair. Neil encouraged me to pursue my proposal and said he would try to attend if the date worked out. I had been reviewing a manuscript for MIT professor David Mindell for his new book *"Digital Apollo."* He and I had the same idea, and it was time to get the old-timers and the new teams to discuss the new challenges. I had progressed in Phoenix with plans with Honeywell, ASU, and U of A. Unfortunately, David and I discovered a conflict with our planned March 3-5, 2008. He and his wife were expecting a new child at that time, and was unable to attend our conference in Tempe Arizona.

Before closing PAT, I was contacted by Roger Launius, NASA's Chief Historian, who was visiting NASA DFRC at the time. At our first meeting in 2003, Roger had suggested that whatever funding I could raise to start a nonprofit would be better put to use to increase his history office budget at NASA. Initially, I had credited his comment to his inexperience as a new Ph.D., but after he came to my home to meet years later, I was satisfied PAT had earned his respect. Our group included John McTigue (a noted DFRC engineering manager and PAT's Chairman of the Board, Bruce Peterson (NASA Lifting Body Test Pilot), and Bill Dana (X-15 Test Pilot) whose wife, Judy, was on the PAT Board).

2008 Go For Lunar Landing Conference

Our efforts firmed up for March 3–5 in Tempe Arizona. The two-day conference consisted of six panels covering broad aspects of the technologies pertinent to the *Constellation's* new lunar lander, the ALTAIR.

Go For Lunar Landing
From Terminal Descent
to Touchdown
Conference

March 4 and 5, 2008
Tempe, Arizona

March 4 - 5 2008 brochure front panel

December 9, 2009, Alatir Conference at JSC

I convinced the NASA Constellation Altair Lunar Lander project office at the Johnson Spacecraft Center to host four moonwalkers for a two-day conference. The last day included the four astronauts, Neil Armstrong, Gene Cernan, John Young, and Harrison (Jack) Schmitt. The highlight of the last day was the translation of Neil and Gene's discussion of the physiological importance of the free flight training provided by the LLTV.

Wide attendance came from almost all the NASA Centers and industry, including the ALTAIR project office from NASA JSC in Houston and the NASA Dryden Flight Research Center at Edwards. Afterward, the ALTAIR project office designated Dryden to lead a trade study to determine options for free-flight, earth-bound training for the new ALTAIR Lunar Lander. I served as a NASA SAGES (Shuttle and Apollo Expert Services) consultant on that team from May 2008 till *Constellation* ground to a halt in the summer of 2010. I also consulted with Orbital Sciences Launch Systems Division in Chandler, AZ, as they formed a team to update the Apollo LLRV gimbaled jet design with current technology. Different team members determined the suitability of contemporary vertical takeoff and landing (VTOL) vehicles, which would run free-flight lunar simulation missions here on Earth to train future astronauts. The trade study concluded the best choice for a new lunar landing training vehicle would be a new gimbaled jet based on the ten feet high (half the height of the LM) Apollo LLRV. The proposed Orbital Sciences new LLRV was 15 feet tall, about 50% of the proposed *Constellation's* ALTAIR.

Neil Armstrong and Gene Cernan

Excerpts From the Video Taped Audio Transcriptions

Neil Armstrong: *"Apollo Eleven's running short on fuel is widely known and has been widely discussed by the press and others, and even in mission control here, people were biting their nails.... I had been flying the machines at Ellington; we took off with nine minutes of jet fuel and someplace between ninety and a hundred seconds of rocket fuel. So we were always landing with twenty, or fifteen or twenty or twenty-five or 30 seconds, we had to – and so it wasn't very much of a concern to me in Apollo Eleven because it was just like I was used to and if I made a mistake it would be a bad one. So I think that was a ... Dave Scott thinks this was one of the more important aspects and one that I haven't really thought much about. He does think that it is an important experience to make yourself have the mindset to be comfortable in a short fuel situation."*

Gene Cernan: *"Neil says it exactly the way I feel; you know to me, it's one step closer to the real world. You are out there at four or five hundred feet. You got five minutes of fuel, fifteen, whatever the hell it is, it is only you and your maker, OK. You're running a problem in an LMS in a simulator; you have a problem, you go topsy into a crater, push the freeze button, let's go on and have a cup of coffee and talk about it. When you are in the environment Neil is talking about, you don't have a freeze button. You are going to either do it or not do it, and I want to know that I can do it and feel comfortable before I get out there two hundred and fifty thousand miles away and have to do it one time successfully. I think putting yourself in that environment, quote, a risk environment, you know John [Young] is right, I don't want to crash an LLTV, but I would rather have the option of screwing up here and learning something and getting out of it and doing it again because I don't have that option when I am landing on the moon. And so it's not ... there is a lot of psychological effect to the comfort level I think we gain when we're in that real world in a lunar module, landing on the surface of the moon that we gained because of the environment we found ourselves in the LLTV."*

(l-r) Cernan, Armstrong, Ottinger, Schmitt

Armstrong, Ottinger, Cernan

On the far side of the table, Armstrong, Cernan, Young, Ottinger

2010's Decade

My NASA SAGES consulting ended the last year or two as our trade study team disbanded after the Constellation program was canceled. However, I continued to lecture senior engineering classes at Utah State University, UC San Diego, and solicit funding for a five-university competition between their capstone engineering courses. The student teams designed, built, and flew scale models of the LLRV getting a head start to enter their new careers.

Spacefest 2017 X-15 Panel

In 2012 I attended a Spacefest conference in Tucson; again in 2017, I participated on an X-15 Panel with Joe Engle. Apollo astronaut and Commander of the second orbital Space Shuttle Mission. Of Twelve X-15 pilots. Joe is the only X-15 pilot still living today.

His X-15 experience started with an account of his first X-15 flight deserving, including recorded from his NASA JSC Oral History. https://historycollection.jsc.nasa.gov/jschistoryportal/history/oral_histories/englejh/englejh_4-22-04.pdf

WRIGHT: Tell us about the few days right before you took your first flight in the X-15.

ENGLE: Oh, total anticipation. I mean, no concerns, no anxieties at all. I do remember going down to NASA [Flight Research Center, Edwards, California]. I had a motor scooter at that time, a little Lambretta motor scooter, and I remember driving across the back desert roads from the housing area down to Dryden [NASA Flight Research Center], and sitting in the simulator, and just kind of like my cockpit in the sandpit back home, just sitting in there and flying and imagining, going through the profile, that first flight profile. Being down there, actually, whenever I could get any of the simulator operators to be there. Even on weekends we'd go down and go through.

I thoroughly enjoyed it. I didn't think I'd every enjoy flying a simulator instead of an airplane, but I did, because I knew it was going to be a demanding airplane to fly. It was a great airplane. It was just a wonderful aerodynamically stable airplane at low speeds, so, really flying and landing it basically was not a real hard task. The fact that you continually operated at the edges of the speed and altitude envelope with the X-15, because that's what it was designed to do; go find out where the edges were, that made it a very intense flying task. But prior to the first flight, it was strictly one of getting as ready as I could be, to not make any mistakes.

It turned out, on the first flight, there was an electrical malfunction that actually took away all the instruments, except the G-meter [gravity meter] and the pressure altimeter and the pressure air speed, which only operate down at lower speeds. So, having gone through those

profiles and just learned them verbatim, knowing what all the different cues were when you came back down, what attitude to hold, because there wasn't any angle of attack, but the nose just a few degrees above the horizon and then when you got there, so many Gs build up, and then push over and then slow down until the air speed worked.

That actually paid off, because that happened on the first flight. Like anything else, if you're ready for it, it's fun. You're looking forward to the next failure. You're looking forward to the simulator operator to give you your next failure. [Laughs]

WRIGHT: During that first flight, you opted to take the aircraft into a slow roll. Could you share with us what led to this maneuver and what was the reaction in the control room?

WRIGHT: You bet. That was something that, quite honestly, I had not anticipated doing. I didn't think one way or the other whether it was the right thing or the wrong thing to do. I was a fighter pilot, and maneuvering in airplanes is something that you just do all the time and think nothing of it. A roll is one of the more benign maneuvers you can do. Because of the other failures, I had been concentrating in the cockpit on the G-meter and watching it come up.

So when I got pushed over and looked out—I had never been that high and that fast before, and I looked down at the lakebed where I was going to land and it looked like it was really under me and passing by in a hurry. Overshooting was one of the things you don't want to do, because you can't get back to the field then. Also, pushing over to a negative angle of attack, where the air is coming from the top of the wing rather than the bottom of the wing, because in that condition, that was one of the conditions that the X-15 became unstable. I had learned that in the simulator by being exposed to that and actually being told, you know, this is one of the things you don't want to do, is get negative angle of attack.

So instead of pushing over to get the nose down into the denser air, I had rolled the airplane, and it was just crisp, it was a beautifully flying airplane—and, I mean, banked it. I just rolled it over and pulled the nose down and let it dish out to get the nose pointed down, heading down into thick air, and got the speed brakes out and it slowed up, and really didn't give it a second thought. Landed. Didn't give it a second thought other than, —Boy, this thing really is a nice flying airplane, and landed.

It came up at a debriefing. One of the engineers came over to me and said, — Hey, you didn't roll that airplane, did you?

I thought he was kidding. I said, — Who, me? He said, — I didn't think so, and it dropped.

And what had happened, they didn't have real good instrumentation in those days either. The roll angle would go out to 90 degrees one way and 90 degrees the other way, so the trace on the oscillograph went out and then started on the other side and came back in. They thought it was a drop out of data at first. I didn't hear anything about it for about a week, and then they developed the film that the little camera that looks out the back of the window and it showed the roll.

That's when I guess Paul [F.] Bikle called [Robert A.] Bob Rushworth, who was the senior X-15 pilot, and told him.

Bob called me and he said, — Did you roll that airplane? I had to think about it. I said, — Yeah, I guess I did. And he said, — Why? And I told him why. [Laughs] And Bob said, — Oh, okay. Well, I'd have done the same thing. Let's go down and explain that to Mr. Bikle. So we went down to see him, and he was very stern about it. He said, —Why did you roll it? I told him why, that I was concerned about overshooting; I didn't want to go negative angle of attack, so I rolled it to dish the nose out. And Paul was a pilot, too, and he said, — Oh, okay. It makes sense to me, but don't do it anymore. He said, — The rest of the guys are going to want to do that and we don't want that. [Laughs]

WRIGHT: It's a good thing you did it on your first flight.

ENGLE: It was a very good thing.

Joe Engle Space Shuttle Commander
from https://en.wikipedia.org/wiki/Joe_Engle

Engle was commander of one of the two crews that flew the Space Shuttle Approach and Landing Test Flights from June through October 1977. The Space Shuttle *Enterprise* was carried to 25,000 feet on top of the Boeing 747 carrier aircraft, and then released for its two-minute glide flight to landing. In this series of flight tests, Engle evaluated the Orbiter handling qualities and landing characteristics, and obtained the stability and control, and performance data in the subsonic flight envelope for the Space Shuttle. He was the backup commander for STS-1, the first orbital test flight of Space Shuttle *Columbia*. Together with pilot Richard Truly he flew as commander on the second flight of the Space Shuttle, STS-2, becoming the last NASA rookie to command a spaceflight until Raja Chari in 2021 on SpaceX Crew-3. He was also mission commander on STS-51-I and logged over 225 hours in space

A New Friend, Angus Rupert, Retired Navy Flight Surgeon

In 2016, I presented a paper on Piloting Spacecraft at the National Space Biomedical Institute (NSBIR) in Houston. The following copy of an e-mail I sent August 31 2021, sums up the value of that experience Begin forwarded message:

C. Wayne Ottinger

From: wayne ottinger <<u>wottinger@aletro.org</u>>
Subject: Wednesday at the museum
Date: August 30, 2021 at 8:41:28 AM MDT
To: "Barry John L." <<u>JBarry@WingsMuseum.org</u>>
Cc: ottinger robert <<u>robert@realitygarage.com</u>>, Rupert Angus <<u>ahrupert@gmail.com</u>>

Angus's mention of the aerospace medical association subgroup meeting reminds me of our September 1916 conference in Houston where I gave an Apollo LLRV/LLTV paper at the predecessor group, NSBIR (National Space Biomedical Institute of Research) That is where I met Angus and the X-15 was the keynote at the dinner on 9/22/16. Here is the PowerPoint presented by I believe John B Clark from the Baylor College of Medicine;

I have some X-15 artifacts and poster boards I could bring if you would like that might be appropriate to have for display Wednesday's visit. Angus and several others in the group spent lots of effort in analyzing this accident and it would be nice to have on this occasion.

I have consulted Angus since I met him:

- Medical Research Scientist
- U.S. Army Aeromedical Research Laboratory
- Mar 2008-Mar 2020 - 12 years 1 month
  Spatial Disorientation, Accident Investigation, Multisensory Cueing, Vestibular Psychophysics, Balance Prostheses

From <u>https://www.onr.navy.mil/en/Media-Center/Press-Releases/2001/Landing-On-His-Feet</u>

For Immediate Release: Jan 01, 2001

Sometimes, good ideas materialize in some very unlikely places. Take spatial perception for instance. Navy Captain Angus Rupert took a recreational parachute jump back in the 70's, and in his free-fall toward the ground realized that even while tumbling he could tell the direction of down just by his sense of touch as the wind pushed against him. Twenty years later, Angus made that idea a reality and incidentally developed a product to help pilots combat spatial disorientation in flight. The Tactile Situational Awareness System (TSAS) is a template of actuators built into a flight vest. The vest creates vibrations on the pilot's torso, and is mapped to points in the environment to confirm the correct information for the pilot on pitch, roll, airspeed and altitude. If the pilot banks to the right, the vibrations are felt on his right side. If he pitches sharply to the left, he feels them on his left. "We orient ourselves by vision, the inner ear, and our somatosensory - or sense-of-touch - system," says Rupert. "If it's pitch dark, vision isn't going to help, and if centrifugal forces are such that we think them gravitational forces, then both our vestibular and somatosensory system have failed us, too, in that they have provided false information." "Severe disorientation can put a pilot very swiftly in an unrecoverable situation," says Dr. David Street, ONR program manager for the TSAS vest. "This is exactly what we need to prevent. "Other uses for the vest are target or threat location in hostile situations, for creating an 'artificial down' in spacecraft (where in weightlessness astronauts also become disoriented), and in prosthetics where persons with severe balance disorders must find a way to stay upright. The TSAS is currently under development at the Naval Aerospace Medical Research Laboratory in Florida. Rupert is a Navy Flight Surgeon assigned to NASA's Johnson Space Center. ONR is looking at providing further funding for an underwater application of the vest.

The Neurovestibular Challenges of Astronauts and Balance Patients: Some Past Countermeasures and Two Alternative Approaches to Elicitation, Assessment and Mitigation
Article
Full-text available

- Nov 2016
- insert Ben D Lawson.jpg Ben D Lawson
- insert Angus Harrison Rupert.jpg Angus Harrison Rupert
- Missing image/leave space Braden J. McGrath

Astronauts and vestibular patients face analogous challenges to orientation function due to adaptive exogenous (weightlessness-induced) or endogenonyergation into the us (pathology-induced) alterations in the processing of acceleration stimuli. Given some neurovestibular similarities between these challenges, both affected groups may benefit from shared research ...

2020's Decade

I lectured and participated in two national aerospace conference panels. The AIAA Sci-Tech 2020 in January and the November 4, 2021 American Control & Guidance Systems Committe (ACGSC) in Sand Diego California.

AIAA SCI- Tech 2020

January 7 2020 in Orlando, I joined a 2 hour panel discussion at this national conference.

The AIAA Scitech 2020 in Orlando

Engineering Apollo Panel: L to R John Tylco MIT, Frank Hughes, Dave Scott, Wayne Ottinger

I told a story just before Dave's response via the following transcript, about the Shuttle Training Aircraft (STA) sustaining the free flight training at much greater cost that the LLTV

Dave Scott's Quotes on Autonomy

One point on your question. There's a trend today towards autonomy, which is great, I mean driverless cars, etcetera, driverless airplanes, etcetera., and I suspect when they get ready to go back to the moon there will be a trend towards an automounous landing. And the Chinese did great, if you look at the TV the Chinese took in their recent lander, right, and it stops, it hovers. It has obstacle avoidance obviously and it picks a nice place to land, brilliant, it's really good. The problem with the human in this, situation is if you have a problem and have to take over, right, you need to be in the loop and the only reason, the best way to be in the loop is to be flying it manually and by that, I mean holding the sticks. the ROB, the LPD or whatever it is, right, you are not manually moving the thrusters or the valves and the thrusters, but you are flying it manually. You're in the loop, you're mentally in the loop. So, you can respond to a problem much more quickly than if you're out of the loop and a red light comes on that says you have a problem.

First thing you have to do is interpret the red light, you have to make a decision as to whether the red light is a real problem or not a real problem. It takes mental time, and there is just no mental time when you are landing on the moon, unless somebody comes up with a better propulsion system or a gaseous propelent or something that doesn't weigh anything, but based on what we know today there is just no time to make a decision if you are surprised by the decision, so I think that the beauty of the LLTV is to put you in the loop you're in the zone, you're in the flow, you know, you're not distracted by anything because you're in the focus and if you do have a problem you're going to pick it up quickly and make whatever decision you have to make to correct the problem, or if you have to abort, abort, but you're in the loop. I think the LLTV teaches you to put you in the loop mentally, which is another comfort zone that you are comfortable about what's going on. I think that's very important when you put humans into a machine.

"(Tylkco) Was there any doubt that you would switch to manual mode for your Landing"

For me, no, I was going to go manual, definitely, just because I was in the loop, right, and wasn't a matter of you know, I'm a big fighter pilot. wah, wah, wah, right, that's not the point, the point was that I had a comfort zone and I knew that I can get it down, you know, with the fuel I had and I could look around. I knew I could do that, and if a problem showed up shows up Jim Erwin was brilliant on systems, golly, he really knew the LM and if a problem shows up, Jim's going to tell me what the problem is instantly, you know, and we'll get it down. That's the point. Though there's no question in my mind, there is one guy who was maybe going to land it automatically, Jim Lovell. Jim was, he told me that much, maybe so, because the computer and the LDC would have landed automatically, right, but, again, it wouldn't have been able to pick out a spot if it had been in a boulder field. But even with obstacle avoidance, which is pretty sophisticated these days, you know I think you need to have the pilot, the guy with the stick, whoever it is, in the loop so that you can get it down, That's the highest reliable system that you can have, with proper training, "(Tylkco) Any other observations Frank?" "(Frank Hughes) Oh no! I agree a hundred percent.

Appendix B begins with Dave Scott's extensive responses to the transcripts I prepared from the NASA video recordings of the 12/9/08 JSC conference hosted by the NASA JSC Altair program office. Dave's in-depth test pilot and LLTV training enabling his successful Apollo 15 landing. Photographs included in the Appendix C displaying the historical period of 1955 - 1975 by the SETP and the publications by Andrew Chaikin constitute an equal value of this book to the stories told.

Epilogue

The Book Text Sources: Approximately 25% of the text is from Appendix C, Dean Grimm's NASA's oral history and Appendix D, Apollo Commander Dave Scott and a few other portions of the book. These pages highlight the 1960s technologies employed by both analog and early digital flight controls emphasizing the need to provide the pilot's ability to be the loop and respond quickly and manually fly the spacecraft in the event an automatic system is flashing a warning light. Bottom line, the Apollo free-flight, earthbound astronaut training was responsible for six successful lunar landings.

Today, the plans to totally depend on automatic, autonomous systems for cars, urban mobility, and low-earth orbit into lunar landings is an unknown risk. The AI driven system planned the first human attempt to land on the moon over a half-century later will encounter known noticeable variable lunar g forces due to the surface geology. Hopefully, no other unknown risks will materialize. Still, the loss of the *"human in the loop"* is now apparently acceptable by NASA's selection of Space X for the first Artemis Lunar Lander. Fortunately, their plans to award another contractor a second contract can give us hope NASA will have a chance to provide a *"human in the loop"* for their final selection.

From: The 1960s/My Move from Civil Service to a Government Contractor:

Retrospect: This significant career move was critical to Apollo as my seasoned experience serving as the NASA LLRV project engineer and the decision to take on the LLTV Technical Director role at Bell. The move allowed me to expand my technical management skills to ensure the transition from flight research to astronaut training was made successfully under critical schedules aggravated by losses of Neil and Joe's ejections, most likely not to have happened had I not made the move.

From The 1960s The 1960s/Apollo Fire (January27, 1967)

"My most formidable management and legal challenge as the Bell LLTV Technical Director" (stopping the two-month delay of training Neil's two days of LLTV training in May 1969, one-month before launch.)

From: Appendix C, close to the end of the 52 pages:

Grimm: *"My most important accomplishment in my career with NASA, per your question here [referring to notes], is I consider in the long haul that the Lunar Landing Training Vehicle project management program management and getting Neil trained in the nick of time so that they could have their FRR and then accomplish lunar landing. That's a significant accomplishment. The second thing would probably have been developing the rendezvous techniques and procedures and training the crews on that.*

9/16/22 Ottinger & 1948 Aeronca Champion Erie CO

On November 4, 2021, at the San Diego Bahia Resort Hotel, I presented a paper at the American Control and Guidance Systems Committee (ACGSC) on the Lunar Landing Research Vehicle (LLRV) Fly-By-Wire (FBW) History, the first aircraft to use FBW on October 30, 1964.

LLRV Fly-By-Wire History
From pages 153 -156 of NASA Monograph SP-2004-4535 (Epilogue)

An element of serendipity exists in all flight research. During the planning phase, and even in the operational stage of a program, it is rarely apparent what findings and contributions will have application beyond the immediate program. Often enough, things emerge that no one knew would surface during the testing, let alone become major contributions down the road. The LLRV program was no exception. Several major contributions to the aerospace community–beyond supporting the Apollo mission stemmed directly from the LLRV program, and radiated out to the larger community. Two specific examples linger in aviation history and technology.

The first is fly-by-wire flight control systems, which led to major advances in both commercial and military aircraft. Neither the LLRV nor the LLTV would have flown successfully without fly-by-wire flight controls. Even if the vehicles' unique performance requirements were achievable with then-conventional mechanical or hydraulic systems, weight requirements would have driven costs far beyond the acceptable.

Initially, the management of the Apollo program was unwilling to admit the need for a free-flight research or training vehicle for the final phase of the lunar landing, which made selling the LLRV program difficult. Happily, the initially cool reception the idea received sent program designers and managers in a new direction, obliging them to develop unproven technology in flight controls.

Ultimately, two forces led the designers into uncharted waters: the need to keep the vehicle's weight and the program's costs down. The result was the analog fly-by-wire flight controls in the LLRV. At that time, the central concern about fly-by-wire systems in piloted aircraft was a lack of confidence in the overall reliability of such systems.

Technologies simply hadn't been developed, let alone demonstrated, that provided the levels of reliability essential for aircraft flight controls. Without wings as a means of generating lift, the LLRV required a very high level of reliability because it's very survival depended entirely on the flight-control system. Loss of flight control, even for seconds, would result in loss of the vehicle, as Armstrong and others found out. Yet concerns associated with a nascent technology were set aside in the face of both need and advantage associated with the new technology.

Extensive experience with the X-15 and other experimental aircraft led the flight- controls research community at the FRC to conclude that conventional hydromechanical systems could not provide the automation and rapid control-surface motions required by higher-performance aircraft. But getting the aerospace community and NASA management to accept this conclusion and fund a research program for developing and demonstrating the technology base needed for practical fly-by-wire development was a different matter entirely.

Flight control research engineers at the FRC viewed the LLRV as a program of opportunity for assessing the performance and characteristics of fly-by-wire technology in a critical application. And over the course of the LLRV program, the performance and reliability of the system proved to be extremely good. Indeed, at no time did the fly-by-wire flight-control system in the LLRV fail completely, and the LLRV pilots' acceptance of, and confidence in, the system's overall reliability helped convince FRC management that fly-by-wire technology was a viable flight-control discipline with applications to future aircraft development

In 1969, while the LLTV was still being used to train Apollo commanders at the MSC, Bikle approved a proposal forwarded to NASA Headquarters for a major fly-by- wire flight research program. A renowned world-class glider pilot and holder of several world titles, including the maximum altitude record for glider aircraft, Bikle was an old-school stick-and-rudder pilot who did not take easily to replacing tried-and-proven methods with unproven new concepts, particularly for something as critical as flight controls. But he also was a visionary who quickly grasped the significance of the LLRV flight controls experience, and he became a strong advocate for major flight research and development of fly-by-wire controls. One measure of this lay in the fact that the proposal for a research program presented to NASA Headquarters for funding had been prepared at Bikle's urging. The program would involve an F-8 testbed aircraft retrofitted with a fly-by-wire flight-control system, using surplus hardware and soft- ware from the Apollo program to provide full fly-by-wire control with no mechanical or hydraulic backup system. Control of the aircraft would depend entirely on the primary fly-by-wire system.

But the program might soon have died had it not been for a fortuitous change at NASA Headquarters. Not long after returning from his trip to the moon, Neil Armstrong had been selected to head up all NASA aeronautical research and development programs. Perhaps better than anyone else at the time, he understood how important fly-by-wire had been to the LLRV. He was also aware of the feasibility, the value, and perhaps even the necessity of applying this technology in future aircraft. And he understood the magnitude of the task of developing the technology to the point where it would become acceptable to the aircraft design community and, especially, to pilots.

Armstrong approved the program–with the strong recommendation that they employ a digital rather than an analog system. His endorsement reflected his experiences with the X-15, with the LLRV and LLTV, and with the total reliability of the Apollo system hardware and software that had taken him to the moon and back. From the confluence of all this came support and funding for the F-8 Digital Fly-by-Wire research program that laid the groundwork in the early 1970s for the acceptance by and application of fly-by-wire technology to modern-day commercial and military aircraft.1 Had it not been for the LLRV's contribution, it is highly **unlikely** that the development and acceptance of fly-by-wire technology would have occurred as soon as it did.

The second long-term impact stemmed from Bikle's decision to "projectize" the LLRV program. Neither easily done nor popular with managers, in retrospect his decision marked the beginning of a new way of operating at the FRC. It evolved into the way many complex flight research projects would be carried out in the future.

For much of the 1950s and early 1960s the FRC essentially had operated as a single-program facility supporting flight research with each successive X-plane.3 Nearly the entire facility was dedicated to the care and feeding of the three research vehicles that comprised the X-15 program, for instance. As X-15 program activity began winding down, it was replaced over a period of time with a number of smaller flight-research projects that competed within the facility for limited resources and workers.

The success of the LLRV program demonstrated that an independent and discrete project could be carried out with a dedicated workforce and resources while not having to compete continually with other organizations for priority and support. Crucial to the success of this approach, of course, was the dedication and resourcefulness of each project manager and technical leader. Another element central to that success was Bikle's delegation of operational responsibility to the program and site managers, whom he considered responsible for the success of the LLRV mission. The result was a freedom of operation in which each individual's talents and ingenuity could be utilized immediately in resolving the complex and critical problems that always seem to arise during the flight-testing of new vehicles.

From the start, Bikle realized that the only way the FRC would have a chance of meeting the schedule requirements set by the Apollo program was to select good people, delegate the appropriate responsibility and resources to them, and then stay out of their way. Bikle's decision to "projectize" the LLRV program ushered in a revised project management and organizational structure that remains a central element of the Dryden Flight Research Center.

Since the LLRV program, the same "projectized" manner of organization has been used in carrying out numerous other flight research projects and activities at the facility. Within this structure, individuals dedicated to a specific project retain their supervisory ties with line managers. This organizational model allows the project team to focus entirely on the technical aspects of system development and flight-test operations without having to deal simultaneously with normal day-to-day supervisory routines and issues.

The process has not only served the Dryden FRC but has been the pattern for other NASA centers beginning in the early 1990s, perhaps when new methods of doing business were being explored as a means of surviving major budget cuts then taking place within the aeronautics community. As it turned out, NASA's new way of doing business is actually the FRC's old way of doing business, dating from the LLRV program at Edwards' South Base.

Drawing on its experience with the LLRV and LLTV

From the Past to the Future

1988, near the end of a symposium titled "Wingless on Luna," Neil Armstrong spoke of the future he saw for mankind on the moon. Tongue-in-cheek, he briefly sketched the outlines of what the future might be like with a community of humans on the moon, sharing the environment with rocket-powered flying machines:

Compact flying machines should have good usability. Cruising altitudes above 200 feet should minimize visibility problems due to dust, but higher altitudes may be required to avoid irritating joggers below. Rocket exhausts are noiseless on Luna, so rocket ports should be immune from noise abatement [law]suits. As soon as the plaintiff's bar has a Lunar section, however, they can be expected to find some basis for complaint.

Once mankind is living in space, the moon will be a valuable source of raw materials-metals, oxygen, and residues from the solar wind. While "the rocket will carry the - brunt of the load" in lunar flying machines, "low gravity and the consequent low divergence rates should make rocket belts more easily flyable than on Earth."

Flying on the moon in the twenty-first century will require the use of rocket attitude controls perhaps different in some design details from those used in the twentieth century, but the applicable laws of physics will be the same. Future designers of flying machines will then benefit from the twentieth century's extensive experience with rocket attitude-control and lift systems. **And the LLRV's groundbreaking accomplishments led the way.** This work is an attempt to preserve a small portion of this technical legacy. As Armstrong wrote, "Someday men will return to the moon. When they do, they are quite likely to need the knowledge, the techniques, and the machine described in this volume."

Appendix A: The SETP History of the First Twenty Years

The Society of Experimental Test Pilots History of the First Twenty Years

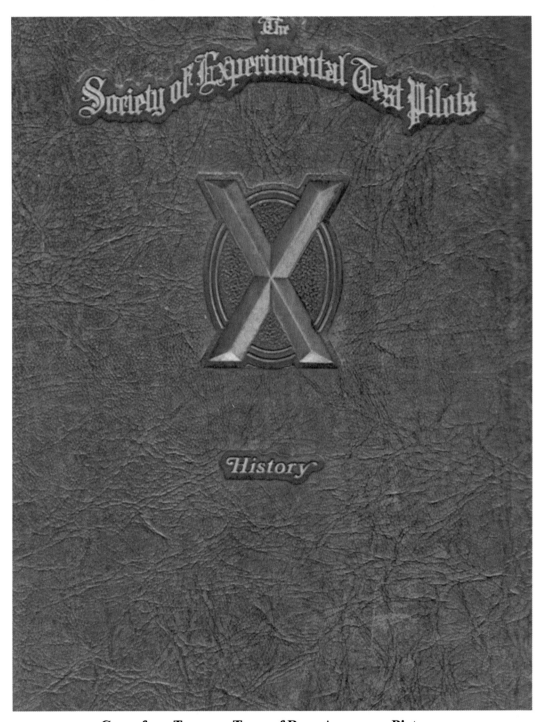

Cover for a Treasure Trove of Rare Aerospace Pictures

THE SOCIETY OF EXPERIMENTAL TEST PILOTS

HISTORY OF THE FIRST
20 YEARS

Taylor Publishing Company
Covina, California

Title Page

Copyright

ACKNOWLEDGEMENTS

The Society of Experimental Test Pilots wishes to express its gratitude to a number of men and women whose assistance in the preparation of this book was indispensable. The presidents of the Society who supported the project during the initiation, writing, and publication, played an important part in bringing this story to you. They were William R. Laidlaw (1976), who urged the Board to budget the money, Irving L. Burrows (1977), who supported the writing, and Thomas C. McMurtry (1978), who pushed ahead with the publishing.

Society members, who laid the groundwork as Chairmen of the Librarian/Historian Committees a number of years ago by writing the initial chapters of SETP's early history, made a substantial contribution. Credit in this category goes to Richard J. Harer, Joseph C. Watts, John J. Shapley, Jr., Richard G. Thomas, and William H. Lawton, Jr.

Relations Committee Chairmen, Robert E. Solliday (1976), Richard T. Gralow (1977) and Stanley P. Butchart (1978), deserve recognition for putting in a year each of their time on the project.

The actual work of researching and writing the History was done by Betty M. Hewson who spent almost three years on the book from start to finish.

To Richard P. Hallion, Jr., Aviation Historian at the Smithsonian Institution, the Society is grateful for his contribution of professional expertise in the field of aviation history while reading the manuscript with a critical eye.

And finally, to the SETP Officers and Section Chairmen who were asked to read and critique the text covering their tenures of office, the Society is especially appreciative. An effort to name all those who assisted in this way would undoubtedly fill a number of pages and to avoid any unintentional omissions, we wish to extend our heartfelt thanks to all who have helped.

IN REMEMBRANCE OF BILLYE

The Society wishes to pay special tribute to the late Billye D. McMains who for eleven years totally dedicated herself to the management of the Society. Not only did she render a wide scope of conscientious service but is remembered also because she was adamant about the need of the Society to document its history.

Billye diligently read the entire manuscript during its writing but passed away on March 21, 1978 just a few months before its publication.

The Society of Experimental Test Pilots reveres the memory of their friend and associate, Billye, who helped them build their fraternity into a world-renowned organization. She will long be remembered for the capable manner in which she conducted Society business and the dignified manner in which she lived her life.

Acknowledgements

FOREWORD

1955, a few civilian test pilots working at Edwards AFB met for lunch at a local cafe. After some stimulated medita-
discussing the idea of forming an organization for the purpose of disseminating information between test pilots
education and safety. Going to work as an experimental test pilot every day is not the best assurance that one
every night. At best, it is a hazardous business. Anything that would reduce the inherent risk was desirable. The
about establishing an organization wherein they could help each other by exchanging information on problems
techniques developed on the job. This, so that each pilot would not have to learn every lesson the hard way. A
sonal experiences in connection with every aspect of flight testing was their objective.

to proceed with the idea. Some very dynamic men solved the organizational problems so remarkably fast that
Society of Experimental Test Pilots was an entity and very shortly staged an awards banquet of high quality.

to have had an Honorary Fellowship conferred on me at the first banquet. Since then almost every person
an important role in aviation or extra-atmospheric exploration has joined as a member and a considerable num-
awarded Honorary Fellowships. With a membership of nearly 1500 test pilots today, it would be almost impossi-
pilot of any major international aircraft or space program without naming a member of the Society. Test flying
business too, and I feel very close to the people in SETP.

nucleus from the desert air base grew in number and stature until they gained international recognition and made
the aerospace world. Governments listen, industry cooperates, the services applaud and laymen stand in awe
pilots perform. They know their flying and they know their aircraft. They also know that their dedication to excel-
tributed to the technological advancement and the welfare of mankind.

Society decided to document its past so we veterans may savor the events of past years in which we had a part
fresh young flyers now coming up may read this History and then want to go out and make their own.

what happened, over the past 20 years, to a Society made up of outstanding test pilots. As it tells that story,
the book will recall to the minds of readers who are SETP members many experiences in their own personal lives
with the Society's activities. It may be a particular test program, a certain technical meeting, an especially enjoy-
and banquet, friends, places, or jobs, that come to mind. It may be the warm feeling of being part of a "tradi-
SETP's first president Ray Tenhoff spoke.

who don't know the Society and are not part of it, here is a fine introduction to a very special group of people
never knew existed. I hope you like the book and I hope you like the people about whom it is written.

J. H. Doolittle

<u>General Doolittle's Foreword</u>

Ryan X-13 VTOL After transition to High Speed Flight

Capt. Iven C. Kincheloe, USAF

Capt. Iven C. Kincheloe, Killed in an F-104 crash July 26, 1958, page 53

Lieutenant General James H. Doolittle, USAF, Honorary Fellow.

Howard Hughes, Honorary Fellow

Vice Admiral Frederick M. Trapnell, USN, Honorary Fellow

Major General Albert Boyd, USAF, Honorary Fellow.

SETP Honorary Fellows

Donald K. Slayton, USAF, Maj. Leroy G. Cooper, Jr., USAF, LCdr. M. Scott Carpenter, USN; Mrs. Dorothy Kinche- ... Col. John H. Glenn, Jr., USMC; Capt. Virgil I. Grissom, USAF; Cdr. Walter M. Schirra, Jr., USN; SETP President Drury ... Jr., and Cdr. Alan B. Shepard, Jr., USN; comprise the Mercury Astronaut Team which was honored at the Sev- ... Annual Awards Banquet (1963) by receiving the Iven C. Kincheloe Award jointly.

Page 123

(l-r): LCol. Fitzhugh L. Fulton, Jr., USAF, Van H. Shepard, Col. Joseph F. Cotton, USAF and Alvin S. White all project pilots on the XB-70 which is pictured in back- ground.

Page 151

U. S. Vice President Hubert H. Humphrey presents the 1...
Kincheloe Award on behalf of The Society of Experir...
Pilots to Milton O. Thompson, NASA Research Pilot.

Page 164

Bob Hope, Page 167

Bell Dual Tandem Ducted Propeller V/STOL (X-22A)

Page 166

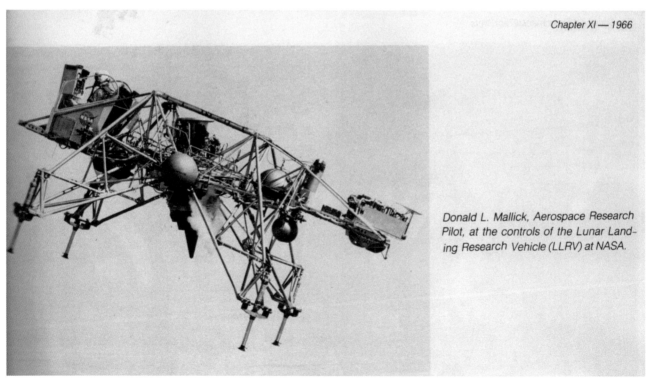

Donald L. Mallick, Aerospace Research Pilot, at the controls of the Lunar Landing Research Vehicle (LLRV) at NASA.

Page 167

Donald F. McCusker, Paraglider Experimental Pilot.

Page 176

Paraglider experiment is executed by Donald F. McCusker for North American Aviation.

Page 176

(l-r): Jack Allavie, Bruce Petersen, Park R. Birdwell, R. L. Stephens and Donald R. Segner at Business Meeting on 12 October 1967 at Sportsmen's Lodge in North Hollywood, California.

Page 193

William J. Knight, Major, USAF, pilots the number two X-15, carrying a dummy ramjet, with no internal air flow, attached to the lower vertical tail. The ship was later prepared for a Mach 6.5 flight.

Page 196

Prime Crew of Tenth Manned Apollo Mission (l-r): Thomas K. Mattingly, II, John W. Young, Charles M. Duke, Jr. received the Iven C. Kincheloe Award in 1972 in the space division.

Apollo 16 lifted off on April 16, 1972.

Page 302

149

KC-135A refueling tests on C-5A with Jesse P. Jacobs and G. Fornell flying the C-5A.

Page 235

150

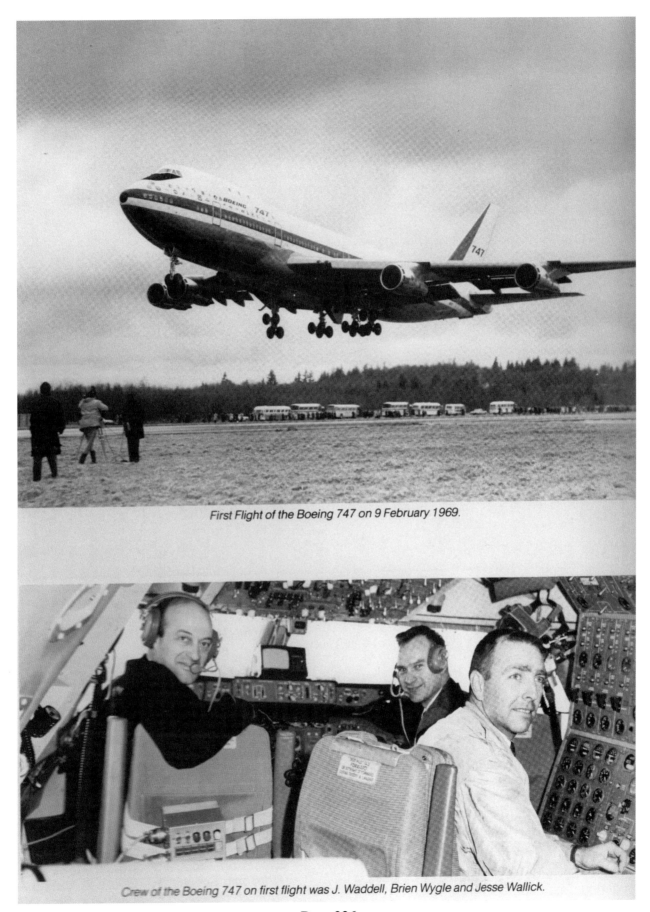

First Flight of the Boeing 747 on 9 February 1969.

Crew of the Boeing 747 on first flight was J. Waddell, Brien Wygle and Jesse Wallick.

Page 236

151

X-15 undergoing inspection by Maj. Pete Knight at NASA Flight Research Center at Edwards, California.

Page 248

On 27 September 1969 the Society held its Annual Awards Banquet. (seated, left to right are): Charles Lindbergh, Mrs. Robert Hoover, Neil Armstrong, Robert Hoover, SETP President, Walt Cloke, VP Public Relations, Jo Doolittle, Lee Atwood, President of North American Rockwell, Mrs. Walt Cloke, Bill Anders, Astronaut, and Mrs. Lee Atwood.

Page 240

Charles A. Lindbergh, 1969 Honorary Fellow.

Page 241

153

Several of the X-15 pilots attended the ceremonies marking the turnover of the X-15 to the Smithsonian Institution in Washington, D.C. In attendance were (l-r): A. Scott Crossfield, Maj. Joe H. Engle, Maj. William J. Knight, Milton O. Thompson, Col. Robert A. Rushworth, John B. McKay and William H. Dana. Unable to attend were Neil A. Armstrong, Capt. Forrest Petersen and Col. Bob White. Two other pilots who flew the X-15 were Joseph A. Walker and Michael Adams, both deceased.

Page 251

French Test Pilot, Jacqueline Auriol, accepts Honorary Membership Certificate from Bob Hoover.

Page 280

(Pictured left to right are): Buzz Aldrin, Neil Armstrong, Dorothy Kincheloe and Mike Collins on the stage as the Apollo 11 crew receive the Kincheloe Award for moon landing the previous year. (1969).

Co-recipient of Kincheloe Award, Darryl G. Greenamyer poses between Red Baron and F8F Bearcat, his two record-getting aircraft.

267

Page 267

155

BGen. Charles Yeager, USAF, is recognized by the Society on the 25th anniversary of his breaking the sound barrier. SETP President Ken Kramer (Lear Siegler) made the presentation at the 16th Awards Banquet in 1972.

Page 300

Hanna Reitsch and Jimmy Doolittle chat during visit to SETP Headquarters in Lancaster, California.

303

Page 303

156

Eastern Air Lines donated the Wings of Man trophy. Scott Crossfield, Vice President of Eastern, made the presentation to Gen. Doolittle at the 20 June 1973 Benefit Ball at the Beverly Hilton Hotel.

318

Page 318

B-1 Supersonic Bomber

Page 324

Called on stage according to their first space flight, the astronauts were introduced individually. Shown (l-r): Deke Slayton, Joseph Kerwin, Ron Evans, Charlie Duke, Al Worden, Ed Mitchell, Stu Roosa, Jack Swigert, Rusty Schweickart, Bob Hope, Pete Conrad, Jim Lovell, Gene Cernan (behind award), Tom Stafford, Buzz Aldrin, Walt Cunningham, Scott Carpenter, Donn Eisele, Gordon Cooper, Alan Shepard, Bill Anders.

Page 330 & 331

John Cochrane and the Concorde Supersonic Jet Transport.

Page 351

SETP President Jim Wood and Billye McMains, Manager, are greeted by German hosts at 7th Annual European Symposium in Munich.

Page 365

Apollo Soyuz Test Project Prime Crewmen: Thomas P. Stafford (standing on left): commander of the American Crew; Cosmonaut Aleksey A. Leonov (standing on right): commander of the Soviet Crew; Astronaut Donald K. Slayton. (seated on left): American docking module pilot; Astronaut Vance D. Brand (seated in center): American command module pilot; and Cosmonaut Valeriy N. Kubasov (seated on right): flight engineer of the Soviet Crew.

Page 370

HISTORICAL PHOTOGRAPHS

Page 379

High flying Lockheed SR-71, LCol. Tom Smith, Chief of Test Force, 1973-75.

Page 381

One of the pilots on the X-24B flight test program was John A. Manke, (pictured) at NASA/Flight Research Center, Edwards Air Force Base, California.

Page 382

Don Mallick flying the non-aerodynamic Lunar Landing Research Vehicle.

Page 383

Bell Aero Systems Tri-Service X-22A

Page 384

Aero spacelines' hefty aircraft is a C-97 (Boeing Stratocruiser) which was modified and expanded to carry large volume cargo. Plane is known as the "Super Guppy."

Page 385

The HL-10 shown here was tested to assist in the development of manned space shuttlecraft capable of maneuvering to ground landings. Four test pilots who have flown this and other experimental lifting bodies are (left to right): Jerauld R. Gentry, Peter C. Hoag, John A. Manke and William H. Dana.

Page 388

Astronauts Fred W. Haise, Jr. (left) commander, and C. Gordon Fullerton, pilot, pictured in front of Orbiter 101 after (ALT) Space Shuttle Approach and Landing Tests. Fullerton gave paper on the tests at the 1975 Symposium.

Page 389

Air Cushion Landing System on the XC-84 demonstrated by Maj. James H. Brahney, USAF.

Page 395

The X-15 changes its image with ablative coating and external tanks.

X-15 Ablative In-Flight

Republic F-105 shown with a conventional bomb load.

F-105 Thunderchief

Appendix B

Photos From Andrew Chakin's Two Books

A Man On The Moon

Cover for Man on the Moon

Emerging from the Apollo 9 Command Module *Gumdrop,* Dave Scott prepares to photograph Rusty Swigert spacewalking 155 Miles above the Earth, who snapped this view while floating outside the Lunar Module *Spider.*

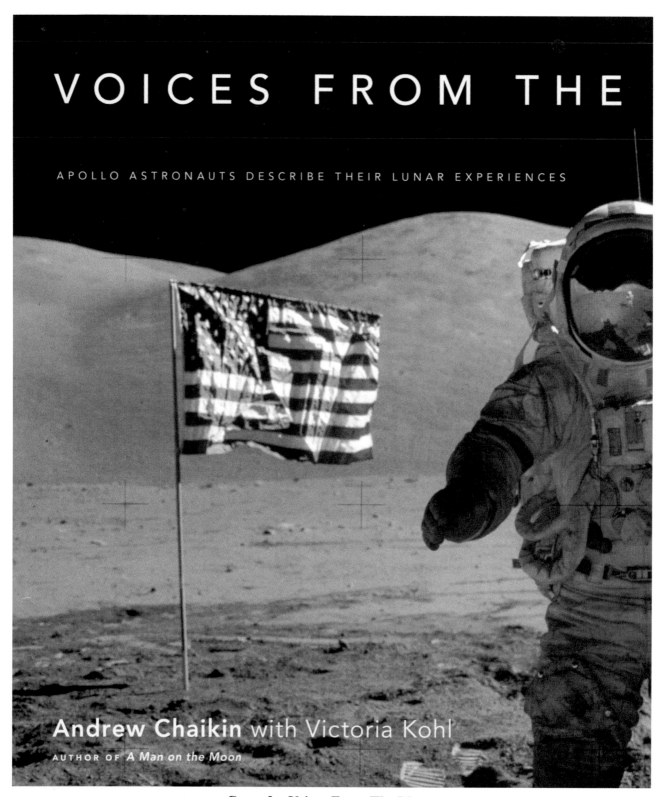

VOICES FROM THE

APOLLO ASTRONAUTS DESCRIBE THEIR LUNAR EXPERIENCES

Andrew Chaikin with Victoria Kohl

AUTHOR OF *A Man on the Moon*

Cover for Voices From The Moon

<u>Dave Scott Training in the LLTV for Apollo 15</u>

Interview excerpts from the book *Voices from the Moon* by Andrew Chaikin with Victoria Kohl
(Interviews were conducted by Andrew Chaikin)

"[The simulator] doesn't do the physical phenomena, and it doesn't put you on the line. The LLTV puts you on the line. Because you know that at five hundred feet, if you screw up, you're dead. And that's good....It makes you focus. In the simulator you can say, 'We gotta run an altitude chamber test this afternoon.' Or, 'I gotta go check my suit. Oh yeah—I'm landing now.' You could let your mind drift. It's no big deal. Flying the LLTV, you don't let your mind drift. So it's good training, 'cause it puts you on the line.

"That thing was a hell of a challenge to fly. Because it's very demanding, and things happen very quickly.... If you want to stop a helicopter, you pull the nose up, and it will stop, and it's floating on the air. In the LLTV, when you pull the nose up, it doesn't float in the air; it falls down....It was very unforgiving, because if you screw up it crashes....But the beauty of it is, it teaches you to run your [mental] computer fast."—Dave Scott

"[The simulator] doesn't do the physical phenomena, and it doesn't put you on the line. The LLTV puts you on the line. Because you know that at five hundred feet, if you screw up, you're dead. And that's good....It makes you focus. In the simulator you can say, 'We gotta run an altitude chamber test this afternoon.' Or, 'I gotta go check my suit. Oh yeah—I'm landing now.' You could let your mind drift. It's no big deal. Flying the LLTV, you don't let your mind drift. So it's good training, 'cause it puts you on the line.

"That thing was a hell of a challenge to fly. Because it's very demanding, and things happen very quickly.... If you want to stop a helicopter, you pull the nose up, and it will stop, and it's floating on the air. In the LLTV, when you pull the nose up, it doesn't float in the air; it falls down....It was very unforgiving, because if you screw up it crashes....But the beauty of it is, it teaches you to run your [mental] computer fast."—Dave Scott

"Landing the lunar module was...the kind of thing where you know that you only have one chance—no two chances, one chance. Everything has to go right. So that puts you right up on the edge of performance....There's an old saying in the program, I don't know whether any of the other guys have mentioned it: 'Get ahead and stay ahead.'...Always stay ahead....So in a lunar landing, it's really thinking ahead. It's planning ahead.... Because if you get surprised, it's going to take away from your time and your mental process. And if you're ahead, you can absorb that....

"As an example: I'm holding onto two [control] handles, and there are a number of buttons in front of me. Now there's a blue button to turn the engine off, and there's a red button to abort. I don't want to push the red button....So when you go into the landing, part of your computer in your mind is concentrating on those three buttons so you don't screw up. In addition to that, you're concentrating on the flying, and you're concentrating on listening to Jim and looking out the window, and the [trajectory]....What if all of a sudden we lose [communications]? What if I can't hear Jim? Then I gotta know what he's doing. So I have to make sure that without him, I can still do the job.

"So you're thinking all the things that you should do, and all the responses to emergencies—you don't get into specific emergencies, but you're just running your [mental] computer as fast as you can run it....And if you get into an emergency situation, things happen so fast, you have such a short period of time, there's no margin. You know—thirty seconds, or whatever it is....You cannot afford to make a mistake....Not so much consciously, but subconsciously, you have all your memory banks running. You're focused entirely on the job, but you're also paying attention.

"So you play all that [in your mind]. 'Cause in the simulations, they've done that to you. They've done all these things to you....That's the beauty of the simulations....In the simulator, you can say, I'm not going to listen to Jim this time. I'm just going to go do it. But in the real world, you have to put everything in the [mental] computer and run it at the same time. So the mental challenge is enormous! I mean, you don't have to focus on all that, but you better. Because if you don't, number one, you could screw up anyway, and number two, if you have a problem, it diverts your attention, and if you're not in parallel processing—and that's what it is, parallel processing all that stuff—if you're not processing everything in parallel at full speed, you're liable to miss something. And if you miss it, either you're dead real quick or you blew the mission. There's no recovery. And you know that, going in. There's one chance, and you've got two or three minutes. One chance and that's all. So, boy, you really tune up for that.

"For that reason, flying the lunar module is a very demanding task. It's the toughest flying job—and I've flown a lot of stuff—the toughest flying job I've ever had." —Dave Scot

<u>Pete Conrad Training in the LLTV for Apollo 12</u>

In the simulator room are, from left, spacecraft technician, Frank Hernandez, Charlie Duke, Ed Mitchell, Fred Haise, and John Young.

"The true Genesis Rock of the Apollo program was the one that I found...the one that was 4.6 billion years old. That was over at the base of South Massif. The question is whether somebody [other than a professional geologist] would have noticed....It was a distinct texture and a distinct color. Again, I can just go back to the question of how...experience enables any professional to filter out the millions of extraneous bits of information and see in micro- seconds what is significant."

—Jack Schmitt

Geologist-astronaut explores a boulder-strewn landscape near Camelot crater during Apollo 17's second moon walk.

It doesn't make any difference whether you're in a suit or whether you're in shirtsleeves. You're still going through the same mental process. It is the mind that you're taking, not the hands... Your mind is not a spacecraft.

—Jack Schmitt

A mile wide and a quarter mile deep, Hadley Ridge looms beyong Dave Scott and the rover during the first Apollo Moon walk.

"Let me give you another analogy. You live in Iowa all your life. And you drive out to Arizona. And you drive up through those trees to the Grand Canyon—blindfolded, okay? And somebody takes you right over to the rim, and then takes the blindfold off, and you look down at the Grand Canyon. That'll blow your mind, won't it?...Well, we drive...up toward St. George [crater], on the first day....You don't have a big peripheral vision. Stop the rover, and get off, and turn around and look, and the goddamn Grand Canyon—Hadley Rille! I mean, that's an absolute mind-blower. Even though you know it's there, but you can't see it, 'cause you're driving this little rover next to the ground. Hadley Rille's over there—you can't see Hadley Rille. You can't even see craters. All of a sudden you get off and you turn around, and there it is! In all its glory. The Grand Canyon of the moon! That's mind-boggling! I mean, that'll blow you out. So I would draw the analogy with the poor little guy from Iowa."

—Dave Scot

Al Worden, Jim Worden, and Dave Scott enjoy the warm Pacific air while awaiting pickup by the recovery helicopter

Drawing on its experience with the LLRV and LLTV,

Dean Grimm earned my respect I have for his technical and management skills, and contributions to all the space programs. He always was fair and personally supportive throughout the good times and the rough times. Here are key excerpts from the pages.

NASA History Office

Dean F. Grimm
Interviewed by Carol Butler
Parker, Colorado – 17 August 2000

Butler: Today is August 17th, 2000. This oral history with Dean Grimm is being conducted for the Johnson Space Center Oral History Project at his home in Parker, Colorado. Carol Butler is the interviewer.

Thank you very much for talking with me today and letting me come out to your home to visit with you.

Grimm: You're quite welcome.

Butler: To begin with, maybe you could tell me a little bit about your background leading up to you becoming involved with NASA and the space program.

Grimm: How far back do you want me to go?

Butler: Maybe how you got interested in aviation and aerospace.

Grimm: I think I was raised on the farm with six brothers and a sister. When I first recall being interested in aviation was prior to World War II and I'd see airplanes flying over our farm every so often and had an interest on what kept them up there and where they were going. Then during the war, we got cards in our cereal boxes about identifying airplanes, and then we'd get little devices sent in other cereal boxes on how to identify enemy airplanes and how to call in and so forth. So that was the very first interest I had in the aerospace industry.

Later, of course, after I graduated from high school I went to the University of Kansas. I originally enrolled in electrical engineering, but the Korean Conflict, as they call it, came around in '50 and I enlisted in the Air Force at that point in time. A lot of my friends were getting drafted, and I decided to choose my own branch of the service. Since I was still interested in aviation, they give you aptitude tests. During the aptitude tests they said I would be a good candidate for an aviation school. I think it was because they had a shortage of aviation candidates.

So I went through a nine-month airplane and engine (A&E) mechanic school in service. Graduating from that school, my grades and aptitude were such that they said I would make a good instructor. So they sent me to a helicopter A&E school and then to an instructor school. Then I instructed in that area for the remainder of my time in the service, except for a period of time when I went through pilot training in the service.

So coming out of the Air Force I was quite interested in conventional airplanes and helicopters and actually in having gained my pilot's license during that period of time, obviously. I came out of the service, and then worked for a short time with Boeing in Wichita on the line, working as a mechanic. I decided that there must be better things in life to do than to work as a mechanic, and so I reentered the University of Kansas after a five-year hiatus.

You asked as a question here, why I picked this field. Well, the reason why I got into aerospace or my aerospace degree was because my background in the service they gave me one-semester's credit. So in engineering language I said, "I are now an aero engineer, rather than electrical." So that kind of explains how I got into that field. Because at that point in time, my age was such that I wanted to get through school as quickly as possible and get some gainful employment, try to make some money.

Upon graduating from K.U., I went to work for Convair, which is General Dynamics [Corporation] now at Fort Worth. Well, it's not even General Dynamics; it's Lockheed Martin [Corporation] now. I went to work for them as an aerospace engineer, flight-test engineer on the first Mach-2 supersonic bomber, the B-58 as a flight-test engineer where I was responsible for doing some of the flight planning and on the various flights on a couple of the airplanes I had assigned to me while they were flight testing.

As the flight-testing program wound down and it transferred over to the Air Force out at Edwards, I wanted to stay in the aerospace fields, so I applied for a position at [The] Boeing [Company] in Seattle. I had worked for G.D. for about a year and a half, I think. So I went to Boeing in Seattle and started in their flight certification area. But because of my interest in flying and the fact that our particular unit developed all of the initial charts and graphs and planning for the experimental flight testing that was being done to certify the first jets for airline use, the pilots over in the flight test area and the FAA [Federal Aviation Administration] pilot in charge of the Boeing certification activities asked me to fly with them and to sit behind them to give them the test points which they had to fly. Then later I would be responsible for reducing that data into certification data, which would then be presented back to the FAA for certifying that airplane for airline use.

So as a result of that I did a lot of experimental flight testing with various airplanes while I was at Boeing, hundreds of hours of actual flying experimental flight testing because the airplanes were experimental up until the point that they were given an FAA certificate. That data, of course, that we generated we used as guarantees to the airlines. They bought the airplanes, and then we did route certification to prove the airplane could do what we said it could do. So that was kind of the background up at Boeing.

After about a year and a half, the work was starting to get repetitive, and I was interested in doing something more challenging so I applied for their supersonic work that they were doing at that time. They were proposing to build a supersonic jet aircraft. This was in '62. Just as Boeing was getting started, [Richard M.] Nixon [?] canceled the supersonic transport project, and I decided that there had to be some things of interest elsewhere.

So I applied at the FAA, Federal Aviation agency, in Oklahoma City, as an instructor, and I taught air carrier inspectors jet certification, air route certification, and all of the things that you do to certify an airline to operate out of airports and for route structure throughout the world. Each one of these air carrier inspectors is assigned to an airline to make sure that the airline follows proper procedures in determining their route structure. So I did that for two and a half years, plus was involved in accident investigations because early in the operation of jets there was a number of accidents.

So from an aerodynamics standpoint and from a certification standpoint, I was involved in that aspect of the accident investigations and actually developed the planning and flew a number of the profiles that duplicated the crashes, except for the final part.

Butler: Must have been hairy at times.

Grimm: It was interesting, because the airplanes at that time, the engines were not what we have now. They were very low-thrust engines, and our thrust rate ratios on those aircraft were very marginal, like 1.01 sometimes. In some cases we had some near crashes while we were doing some of the duplications of the accidents to actually see what was happening. This happened also at Boeing as well as at the FAA.

During the later part of my stay with the FAA, the Langley [Space] Task [Group] force had moved to Houston. NASA Headquarters made a decision to enlist the help of LBJ, [Vice President] Lyndon Baines Johnson, to move the Space Center there. One of my cohorts who had known me both at Boeing and at the FAA had transferred to NASA from Oklahoma City, and he called me and said they had some job openings there that he thought I might be interested in. So I rented an airplane and I flew down to Hobby [Airport] and went out to NASA and interviewed several people and was hired on the spot in a group which was called [the] Operation Support Section, which doesn't have much meaning as far as function is concerned.

But that particular section was set up to have a number of people who had piloting for flight-testing backgrounds to come in and actually work with the various contractors and NASA in flying all of their simulators and developing procedures that we were going to use for Mercury, Gemini and later on Apollo and Shuttle. So there was about six of us, and we went to LTV [Ling-Temco-Vought, Inc.]—we went to Grumman [Aircraft Engineering Corporation]—we went to McDonnell Douglas [Corporation], we went to [North American] Rockwell [Corporation], we went to the Navy, we went to the Air Force facilities to evaluate and to fly their simulations to determine which ones were good, which ones were not so good and also to start developing procedures for the various parts of each flight segment that were involved in, initially Mercury, and then later of course Gemini, the various types of missions we were involved in. In many cases, inputs to the design of the cockpits, control displays, navigation aids, software, hardware, instrument scaling and those sorts of things.

One of my very first tasks when I went to Houston? and we were not out at NASA at that time, our division flight crew operations division was in the Franklin Apartments there in the south part of Houston. My first job was to take the Apollo reference mission, at that time, which was a large stack of books and to look at the lunar part of the mission. At that time, even though we didn't know what the vehicle was going to be or look like, that I was to review the mission profiles and determine what kind of instruments that this lunar landing device needed for the translunar part, as well as the descent and the ascent and then to provide a scaling for all of the instruments that would be needed for that flight.

Butler: That must have been challenging, as you said, not knowing about the spacecraft.

Grimm: That's right, but an interesting characteristic of my personality is that I have never, ever had a job or taken a job where I felt threatened. I always felt at ease because I said that if there was somebody in this job before me, I was sure I was just as smart as he was and that I could learn the job. If there had not been anybody in this job prior to me, then I had the responsibility to learn as much as I could to do that job.

I've always taken that approach on every job I had. It's always a challenge to me, and I enjoy those kinds of challenges. Some people get very nervous when put in that position, but I've never, ever been nervous in those kinds of positions. I love that kind of a challenge.

Butler: It's always good to be challenged.

Grimm: I offered jobs after I got to NASA to a number of friends that had similar experience to me, and they said that the job that I had was too stressful. Of course the job was stressful. Everybody at that time at NASA had stressful jobs, but some people accept the challenge and then some people don't.

Butler: Each person is different.

How did you determine what instruments? What were some of the factors that you looked at to determine what would be needed?

Grimm: You're asking me something that's almost forty years ago. But when you look at the reference mission and see the speeds that the vehicle is traveling after it separates from the command module?I said after it separates from the command module, but at that time, it wasn't decided that this vehicle was going to separate from the command module. At one time we had a vehicle [design] that was called a Nova, which was about 500 feet tall, versus the Saturn V, which was 300 and some feet tall. The part that separated in Earth orbit went to the Moon, landed on the Moon, and came off the Moon and came back and then separated from the part that was going to land back in the water. So you had to look at what the characteristics of a particular part of the flight were and look at the velocities and look at your rates and look at your control characteristics and thrust levels that you had to maintain and what types of instruments that you would need to not only evaluate the situation that you're in but be able to control that situation at the same time. That's about as near as I can explain to you what we looked at, because there were very few numbers in this reference mission.

So after that project was completed, I and basically one other individual by the name of Ed Smith and Hebert Edward Smith was the person who had enticed me to NASA and then who I was with him in his group. We worked for a Captain

Brickel who was the section head at that time, who is now a retired three-star general. I think he's up at Fairfax, Virginia, and has his own consulting firm as all those retired generals do.

So we ended up going to all these places that I previously mentioned, flying various types of simulations that the contractors had developed, determining whether those were valid. We looked at the visuals, we looked at the instruments, we looked at their procedures, to see how these things might be melded together to actually perform an operational function. Once we determined that there was an operational function to be performed with those procedures or with the simulator, then we would recommend that the crews then use those devices to gain some familiarity with what they were going to be expected to do in the future.

As they got closer to flight, we picked the systems that we thought provided the best capability. Of course the astronauts were quite capable in their own right of evaluating these systems and then we'd use those systems for functional training for the flights.

Butler: Would you make recommendations based on these different systems around the country devices that NASA should then build?

Grimm: Oh, yes, oh, yes. We'd also tell the contractors what we felt needed to be done to improve their simulations.

Can you tell if I'm recording or not?

Butler: I believe you are.

Grimm: Okay. I was just wondering here.

Butler: Yes. Everything sounds good.

Grimm: I didn't want to be doing all this blabbing here and not have it recorded.

Butler: Oh, yes. The good thing about this [recorder] is it actually has two different reads, for both of our mikes on it.

Grimm: Now the next question I had is do you want me just to go right on through to the projects that I did?

Butler: If that flows well for you.

Grimm: Are you interested in those sorts of things?

Butler: Oh, absolutely, absolutely.

Grimm: Okay. Because I could jump to major things like LLTV or the Gemini rendezvous or the first Shuttle payload that we developed in my division or a number of things like that.

Then subsequent to that in applying these simulations, I started looking at the Gemini docking because we knew that one of the requirements for the Apollo Program was going to be a rendezvous and docking somehow and someplace. So Gemini was put into the program specifically to validate the capability to be able rendezvous two vehicles in space and to physically dock.

At that time, there was a number of issues. One, we didn't have any rendezvous procedures that worked. Two, we had a docking training vehicle down in, I think it's 227, the big tall building, I can't remember now. I think it's 227, 225, down on the north end of JSC. The whole building inside is painted black, and we had the Agena on rails that would go forward and aft. Then we had a Gemini cockpit that would go left and right, up and down and pitch and roll and yaw.

One of the things that I did was to do a handling quality study with that docking trainer because at that time we hadn't really decided what characteristics we needed in the control system to allow us to control the Gemini space craft adequately and to dock with the Agena vehicle, which was a passive vehicle, and stabilize with that.

During this exercise, we did a handling quality study and I used Wally [Walter M.] Schirra [Jr.] and Gus [Virgil I.] Grissom and several others of the first seven astronauts including myself and one of our other pilots in our group in our section. We actually did a handling quality study and used the Coopers Rating System, which is a system that pilots use to put numbers on a system on how well you can control a device in certain—as you vary the characteristics of a control system, you get a different feel for how the vehicle is controlled, and that's called a Coopers Rating, and you can just assign numbers to it. You always like to have high numbers. Anything above a five is nice. Anything down towards zero is unstable.

So we did a handling quality study there to define rate command, attitude, hold, direct, breakout forces, the jet logic, cycling of the jets. I can't think of the word I want to use, but it's the on-off characteristics. I guess maybe it's cycling. Anyway, we defined the characteristics that we felt were good and then we provided that to the manufacturer to put into the software and to the hardware because the hardware was adjustable and the software could be changed as well.

As a part of that study, I found that first of all the docking light? we had an acquisition light which was a flashing light on the Agena, which you could see from far off. It's like an aircraft flashing light that you see at night flashing. But once you got up closer to the vehicle, they had a light in the cone that lit up the cone, but in space that cone was so bright that it just blinded you with the reflection. We had a COAS optical sight. The COAS means Crew Optical Alignment Sight, which is an infinity focus device where you look through it but the reticle pattern that's inside of it is transposed to the target so that the target and reticle pattern look like they are superimposed on one another in the infinity focus device. That way there's no parallax and your eye doesn't go from looking at the actual thing right in front of you to the target, which was the Agena way out in front of you. They all look like they are at the same point.

It turned out that the COAS was not designed properly so I took it upon myself to redesign, with the manufacturer, the COAS and the reticle pattern and the lighting in the [Agena] cone so that the crew? we redesigned it so that the crew did not see an apparent difference in the brightness between the reticle pattern and the vehicle he was looking at. We also adjusted the lighting on the Agena so that we didn't have a glare there.

Then the last thing I did was there was no reference to the vehicle, and you couldn't tell whether the vehicle was pointed up, pointed down, pointed sideways. All you could see was the cones. So I recommended to my division chief, Warren [J.] North, who I have a great deal of respect for. He's kind of one of my heroes at NASA, always was. I don't know whether you've interviewed him or not.

Butler: We have.

Grimm: He's a very interesting individual, and I love him to death. He would probably be embarrassed if he heard me say that.

Butler: We had a good oral history with him.

Grimm: Good. I'm sure he had a lot to offer. If he didn't, then you were shorted.

But I told him that we needed something on there like running lights that we have on aircraft, and he agreed with me. So I then instituted a lighting study. I had a mockup built of the Agena and we put it on a trailer and we picked the time before the new moon when we had no moon on the backside of NASA. There used to be a road down there along that fence, but I think there's a highway there now. I and another engineer went down there at night, and I put little lights with LEDs out off of a computer, an old analog computer. Took those lights and put them at the appropriate places on the vehicle, front and back. I put green on the right and red on the left and orange on the bottom, front and back, so that you could not only see the orientation of the vehicle, but be able to as you were coming up and rendezvousing with it in the real case that you could tell what your orientation was to the vehicle and be able to know your velocities out and then have a perspective in terms of depth or distance from the vehicle so you could actually slow your approach speed down to a reasonable level so that you could come down and actually engage the docking cone and dock. So that was another little project that I had that I enjoyed.

Actually I did the same thing again on Apollo, because we had [another] COAS [with a different problem] and COAS problems. In addition to that, we didn't have a good docking target. So I worked with Grumman initially and then with [North American on] Apollo to develop a?and I actually did the design of that docking target that we used on Apollo, the target itself and the same amount of work on the COAS so that the crews could accomplish the same thing, the docking part and the rendezvous part. My lighting, my little lights, were carried over, except in this case we got a little more sophisticated. We had a contractor this time, used little LEDs, and because if you put the same power into an LED, red, green and yellow have different intensities just because of the spectrum. To tell you the honest truth, I've forgotten which is which now, but at the time I knew.

They put the proper number of LEDs in each color segment so that each one of them would have the same intensity when looked at by the crew. We put those on Apollo. As a matter of fact, on that docking study back on Gemini, I found that we needed a light when we got up close on the Gemini. So I came up with a light that almost looked like an old fender light off of a '37 Dodge, and I forget what it actually came off of. I think it was a wingtip light off of an aircraft that was modified and put on the top of the Gemini OMS [Orbital Maneuvering System] system, not the Gemini itself, but the OMS capsule behind it that it was attached to. Put it right up over the pilot's head so that when we got close to the Agena that he could turn that light on have an even better perspective.

That worked out quite well. The little running lights, the docking lights, the overhead lights, worked on Gemini so well that we did the same thing on Apollo and developed these lights and had them installed on Apollo because as it turned out the command module [CM] was going to be the passive vehicle in orbiting the Moon and the LM [lunar module] was going to be the active one. So with those lights on there, and then if need be in an emergency, the pilot in the command module could, with these lights and so forth, dock with the lunar module because the lunar module didn't have lights on it as I recall. Now I could be mistaken about that. That's been so long ago that we put lights on there, but I don't think we did.

Butler: That's something we could look into.

That's interesting. There's so many pieces of all of this that had to come together to make everything work. Lights wouldn't be a normal thing that you could think of right offhand as a critical requirement for going to the Moon, but yet it did play such a big role in the docking and making it successful.

Grimm: All these little things add up.

So I think that was one of the things. Of course, one of the interesting things I found out about it is where Rockwell had put the target on the (?) or the target was put on the (?) they actually put a target in the lunar module, I mean in the command module, so that the lunar module could dock. It turned out that somebody had miscalculated the angles and so if they were lined up exactly, the COAS, the reticle, with the?I don't know if you've seen the target.

Have you seen the target?

Butler: Yes.

Grimm: If you've seen the target, if they were lined up perfectly and they tried to dock, the docking cones were misaligned. So it took me quite a while to convince a certain contractor that they needed to correct this. So I actually proved to them that they were wrong, and finally we did reorient the targets so that we did have a match.

Butler: Very important consideration.

Grimm: Right. So that we could actually get a hard dock and latch.

During this time, Captain Brickle had left, and I was given the section by Warren North. When I came to NASA, one of the first, along with some of these other things I was doing, the LLRV [Lunar Landing Research Vehicle] project was coming into being between MSC [Manned Spacecraft Center, Houston, Texas], now JSC [Johnson Space Center], and FRC [Flight Research Center, Edwards AFB, California] now Dryden [Research Center], and [NASA] Headquarters and Bell Aero Systems Company. I'm not sure how I was picked, because there were several other people that had similar backgrounds to mine, flight-test background, airplanes certification, and so forth.

But in any case I was given the project and said it's mine and I had it from almost the time I came to NASA within three or for months until '69, from '63 to '69. I was the program manager during that period of time for the RV. Then the RV was passed off to the operational group at Ellington [Field, Houston, Texas] after I had set everything up, and then I was the program manager on the LLTV [Lunar Landing Training Vehicle]. Then we had the accident with Neil [A.] Armstrong, and then I was brought back into the picture after that. I had the program again totally until right after the lunar landing. But that's another subject. I'm just saying that was another task I had.

Butler: Actually, while we're on that, maybe we could talk a little bit about some of the details of the LLRV and TV. What were some of the differences between the two?

Grimm: There was a lot of differences between the two. They looked almost identical. If you were to go out to Dryden and look at the FRC, or the LLRV they have hanging up out there, and I guess the other one [LLTV] is in the lobby of Building 2.

Butler: Building 2, right there by the Teague Auditorium.

Grimm: You'd think the vehicles were almost the same. The structure looks the same, but it's different. The cockpit is totally different. The engine has changed. The avionics were completely changed. The amount of actual lift rockets on the vehicle from the RV to TV were changed. All of the instruments were different. The hand controller and the T-handle were from the lunar module and not from model shop rework, because those were Gemini controllers that I got, actual ones for the spacecraft that were flew on.

As I told the gentleman here the other day that was interviewing me for the LLTV because he's writing a book on the LLRV/TV, I told him I either lied then in '71 or I'm lying now, and I don't know which, because I told him the hand controllers that were used on the LLRV 1 and 2 were Gemini VI and VII flight controllers. In my interview with [Ivan D.] Ertel, I said they were off of VII and VIII flight vehicles, which that's probably more correct since that was very shortly after the program that I did that interview with Ertel.

Butler: I'm sure that's something that we can find in the records somewhere.

Grimm: The ejection seat was upgraded. We put a top on the vehicle and then later cut a hole in it. Our avionics were completely different in terms of their functionality. The jet logic? the RV was designed basically by FRC to conduct handing qualities, to determine how a nonaerodynamic fly by wire vehicle could be controlled. As a result, their main emphasis was on handling qualities and with some considerable amount of thought on how those handling qualities could be transferred to a lunar vehicle if there were going to be one, which at that time hadn't been defined.

So they built in some variability. So they basically wanted to do a variable stability handling quality study, which they actually did. But as we got further downstream, the Apollo Program translated from a Nova type vehicle to an Apollo type vehicle with a lunar module or a LM as they called it at that time. As a result, we knew that it was going to be a smaller vehicle, and that it was going to separate from the Apollo command module in lunar orbit and that it had to have certain handling qualities to go down to the lunar surface and certain handling qualities to come back because it was being staged from the descent stage.

As a result of the initial definition as a lunar module, and then I began to get some appreciation for the masses of the LM, and it's moment characteristics, if you're familiar with that. We're talking about the inertias, the moment of inertias, about each one of the axis, the pitch, roll, and the yaw axis. Once you gain an appreciation for that and you're trying to design a vehicle to train the crews in 1G gravity, to have that vehicle fly like it's in 1/6G gravity, which you'd be in and around the Moon, and then to have this vehicle that was going to fly in Earth's atmosphere fly with the same control characteristics? at that time we didn't know what all those characteristics were going to be. So I had to build even more variability into the control system than FRC had.

Of course the lunar module had fore and aft RCS [Reaction Control System] jets, one right up in front, one in the back and one on each side. The LLTV, and the RV in that case, had it on left right on each side of the cockpit in front and left right on each side in the back. So we had to devise a different jet logic to control the vehicle, even though their thrusters on the LM were fore and aft and off to the side and ours were here and here. So we're off forty-five degrees in terms of when you fire a system, you've either got to fire in pairs, where they fired front and back, we'd fire two off over here and two off over here [Grimm gestures] to give you the same pitching roll as an example. Then we put a set of backup thrusters on just in case we needed those or needed the extra thrust in case of a problem in training.

So those were the major differences. We had differences in tankage. We had differences in the engine. We had differences in the Doppler radar, radar altimeter, our instruments. Ejection seat structure was something different. Avionics were certainly different. Jet logic was different. Of course since we were using the LM hand controller and the modified T-handle, represented the thrust and weight control for the descent engine, we ended up having to have all of those characteristics in our avionics that would allow the crew to control the vehicle as if he were in the lunar module in the 1G environment.

In order to do this, we had various modes on the LLRV/TV that would allow the engine to gimble. We had an auto throttle on it so that we'd weigh the vehicle at a thousand feet in the air and we'd stabilize it there. We actually had a weighing capability to weigh it itself. Once it weighed itself, it would throttle the engine to 5/6ths the weight of the vehicle.

We had drag compensation so as we started going forward it would take out the drag of the vehicle to make it as if it were flying in a vacuum, as it would on the Moon. Then it would go into the lunar simulation mode where the vehicle would accelerate up to a certain speed, go into the lunar simulation mode, and then the engine would start gimbling and then the crew would be using two 500-pound rockets on each side of the engine gimble to represent the descent engine. Then he would control that descent engine with his T-throttle and control the attitude with the actual LM controller that I had modified and put in the vehicle.

He was looking at instruments at approximately the same angle as he was looking at them in the LM because I had the instrument pedestal moved over. Of course, the RV had a collective stick, a center stick and rudder pedals, initially. I finally had them get rid of that and do the last part of their studies with the Gemini controller that they'd modified that I'd gotten from either Gemini VII or VIII and a modified T-handle for the descent rockets that we had on the vehicle.

Then later, of course, we had a Pitot tube probe with the wind direction and we had an anemometer on top for velocity because of the accident, I should say, that caused that.

So when you get down to it, there are more differences between the vehicles than similarities. The only similarity is that from an uninitiated viewpoint they look the same. Does that give you a (?).

Butler: Oh, absolutely. That's very interesting that they were so different, because like you said people do assume that there are so many similarities.

Grimm: See, the FRC was doing one thing, which was great because it gave me initial data on the vehicle. But then I had to do something else because I was told to make a vehicle that flew like the lunar module.

Butler: Quite a challenging bit of engineering and planning.

Grimm: In the whole program there was only two people from NASA on this program, I and one other guy.

During this time, I had either the operation support section or the operation support office, I forget what it was called. The operation support office, and then I had other people that had teams that I supported the crews with. We did the man rating on the two altitude chambers, the first crews that went in there, and actually man rated that chamber with Apollo spacecraft and the lunar module. We had responsibility for the design of the crew station on the command module, and then I had an assistant manager that was responsible for the design of the crew station on the lunar module.

Then at the same time, I had responsibility for the neutral buoyancy facility, the air bearing system, and all of the mockups and trainers. So maybe that's digressing a little bit.

Butler: It's good because it shows you had so many things you were focusing on at the same time and each of them had different levels of importance and all played a very vital role.

Grimm: They were all different. Very different. So we sort of jumped there.

I think that one of your questions here was discuss my work with the operation support section. I think I pretty well covered that in terms of flying the simulations, the lighting studies, the mockups, the docking and rendezvous things, the COAS study, the handling qualities study, defining the scaling of the instruments for the lunar module, and at the same time I had the LLRV/TV project and I was the program manager during that. I was responsible for the budgeting, for the direction of the program, for getting the vehicles built, for the facilities that we ended up with at Ellington.

Butler: What were some of the challenges? Obviously, budgeting would have been a challenge at the time. But what were some of the receptivity from astronauts and from others around NASA?

Grimm: I think the astronauts initially weren't involved in it. I think the big pusher on this was Dick [Richard E.] Day, who was my assistant division chief boss at the time, and Warren North and [Robert R.] Gilruth. When FRC, Flight Research Center, and Paul [F.] Bikle came, who was the director out there at the time, came to talk to Gilruth about this after they'd received Textron's, which is now Bell [Aircraft Corporation], proposal, or vice versa, I'm not sure. Textron, I guess, owns Bell.

But [NASA] Headquarters had seen the proposal and they said, "We don't know what to do with it." They sent it to FRC. FRC liked it, but Headquarters wouldn't give them any money because it didn't pertain to the program. So they came to MSC or JSC at the time, and Warren was interested in it. Neil Armstrong had been selected shortly after that time. Gilruth and FRC convinced, maybe not convinced but saw a mutual interest there with JSC. So JSC then, Warren and Gilruth went to Headquarters and said, "We'd like to have some study money to have Bell study this more and define the vehicle."

Originally the pilot was sitting on top of the engine. There was no cockpit sitting out in front, and the avionics were? in other words, this was a pyramid. After a while, in figuring out where CG [center of gravity] was, this vehicle flew better upside down than it did right side up. So to get everything back in a proper perspective, Bell moved the cockpit out in front and all the avionics and some tankage to the back to balance the CG. Then even there, we had to put weights on the legs in various positions to balance the crew weight and the vehicle because the crews weigh differently. If you took somebody like Neil [Armstrong] and somebody like Jim [James A.] Lovell, there's probably thirty, forty pounds difference in weight there, this vehicle was very sensitive to that.

So we either move the avionics to the back of the vehicle, back into the side, we could move forward and aft, and left right, and if we couldn't get all of the adjustment to keep the CG in the right spot, because it had to be in about a one-inch spot, that we put weights on the legs to do that with shot bags.

I think I digressed here, but, from some question.

Butler: We were just talking about some of the challenges and some of the reactions to the (?).

Grimm: Oh, as far as what their thoughts were, I think once they started seeing the concept of the vehicle, and then having the vehicle built, and then seeing its characteristics, one of the interesting things that came out of the FRC flights and the studies were that they had almost no feeling for the attitude of the vehicle.

You fly a conventional aircraft, you have reference with the nose and the wings and so forth in terms of your attitude and to some extent your deceleration rate and you descent rate and so forth as you come in to land. You fly a helicopter, and you still have a fairly good visual reference, and less, maybe, of a pitch orientation with a helicopter, especially if you are sitting right out in front on some of the helicopters. Some helicopters that have a long nose, you would have a better perspective of pitch. A helicopter has a much more apparent angle as it's approaching a flare and then landing vertically, which is what we are going to have to do on the lunar surface, because we can't just go coasting along like we would on an airplane and put our wheels down and land. We might end up in a crater, we might end up on top of a boulder, so we have to basically come down vertically, and know all of our velocities and then the last ten or fifteen feet come down vertically knowing all the residual velocities.

One of the things that they found out very quickly was that in flying the RV out, when they were in the lunar simulation mode and they were buzzing along as if they were over the lunar landscape in a descent and they decided there's a spot that I want to land at, the first thing that happened to them is that they were long past that spot before they could get stopped. The reason why is that you have to have six times the attitude here. You're one-sixth gravity, so you have to have approximately six times the attitude in terms of pitch, with these rockets firing to decelerate you and to null out that forward velocity.

The pitch attitude was very hard to define by just looking out the window of the LLRV and it was on the lunar module as well. That's one of the first things that Neil came and Pete [Charles Conrad, Jr.], both came back and said, "It's good that we have the LLRV/TV to fly because we would not have appreciated how high we had to pitch to null out forward velocities in order to sit down in a particular spot," because you had to think a long ways away and you had to start the action a long time. Then you had to wait a long time for the deceleration to occur. If you didn't have the pitch attitude, you just had to keep cranking it up to null out that velocity.

Butler: It must have been rewarding for you to see that it was so useful.

Grimm: Yes. Neil said that. It was very comforting to find out that the lunar module flew quite similar in terms of attitude, in terms of handling characteristics as the lunar landing training vehicle. Of course, that's one of the satisfactions that you get personally as a reward other than some little medal that somebody might pin on your chest and say "Good boy, Dean." It's something that you remember long after you're gone from the program. It's something you feel very proud of that you did even if nobody else knows, or very few people know that you did it.

Butler: It certainly is something to be proud of.

Grimm: I don't know whether I've covered your question or not.

Butler: You've covered it pretty well.

Grimm: There are a lot of other subtle differences between the vehicles as well, but I think those were the major things. The major thing that was important to the astronaut was the recognition of the attitude and that they had to refer to instruments rather than look out the window because you really couldn't judge your pitch by looking out the window. The second thing was to crank up a good pitch angle and hold it and wait to see what the response was as you were descending to the lunar surface to null out your velocity.

In Neil's case, he got down fairly close to the surface and was horizontal, stopped his descent and flew along horizontal just above the place where he was getting the dust cloud because he could see places where he didn't want to land since we was horizontal now. Then rather than have to pitch the vehicle up, it wasn't much of a pitch then to stop the vehicle because he was only going forward at a few feet per second at that point in time. So he could pitch it up at a relatively benign angle and stop his velocity and then set down, which he did with maybe ten seconds' worth of fuel left.

Butler: Did you follow the landing very closely? Do you remember where you were?

Grimm: Oh, yes. I wasn't there. I had told Deke [Donald K. Slayton] that since I had spent the last two years prior to the lunar landing almost totally dedicated to that vehicle, plus some of these other tasks that I had, I was usually out there twenty hours a day, seven days a week for two years because there were always problems and there were always decisions to be made. We had accidents and we had to put a third vehicle in the wind tunnel to determine what our real problems were. There were always electrical and electronic bugs to fix, things that we couldn't imagine happening on the vehicle but it did; such as, the system always switching into backup at certain times when we were flying it. We had no idea why until we finally wondered if that big search radar that Ellington had was causing us a problem. So we had telecommunications division come out and set up a system to measure the electromagnetic pulse we were getting from the search radar. Sure enough, when that thing was pointed right at us, in flight or on the ground, it would switch our electronic system.

If you believe the readings, everybody out there was sterile.

Butler: Well, that's not quite so good.

Grimm: Because the energy was so strong. What we got the Air Force to do with some cajoling was to turn the radar system off or point it in a different direction and leave our sector clear while we were flying the crews out there. But that jumps ahead quite a ways from your list of questions, if we want to take things in order.

Butler: Sure. We can go back. It all seemed to flow logically there, so that's all right.

Grimm: There's probably some other things we could talk about on the project, but—

Butler: Sure. Well, we can always come back to it, too.

Grimm: You have my orbital rendezvous work with Buzz [Edwin E.] Aldrin [Jr.]. That's a long story. The one thing I'll say about Buzz is that he could come up with more ideas in five minutes than would take me twenty-four hours to shoot down.

He had a lot of good ideas, and some we incorporated into the orbital rendezvous procedures. Buzz wrote his Ph.D. thesis on orbital rendezvous when he at MIT [Massachusetts Institute of Technology] before he got selected as an astronaut. So that's why they always called him Mr. Rendezvous, if you've ever heard that statement.

It's interesting that Buzz, I don't know how widely he admitted this, but he admitted that after working on the actual orbital rendezvous with me and others during the Gemini spacecraft flight, that he found out that his thesis was wrong, or at least some parts of it were wrong. [Laughter]

Butler: It's always a learning process.

Grimm: I'm sure it is. I'm sure a lot of those Ph.D. thesis aren't as meaningful as probably the one he was working on, too. But it's an interesting story about how I got to be working on orbital rendezvous.

I had been flying simulators, and at that time I was flying a number of simulators, both at JSC and elsewhere, on the lunar module simulators or simulators duplicating lunar module characteristics for rendezvous. It was a very crude approach that a number of people were using for rendezvous, and you could maybe only rendezvous maybe once in twenty times and then it was by brute force if you had enough thrust.

On the Gemini Program, it was mandatory that you have an orbital rendezvous on Gemini and if we were going to have two vehicles separate and get back together again on the Apollo Program for the lunar landing. We were six months away from Gemini VI and we didn't have any orbital rendezvous procedures that worked. The simulator there at JSC cratered every time that it tried to rendezvous. Another one at JSC worked part of the time. There weren't any manual backup procedures to ensure rendezvous if the automatic system had errors in it or did not work.

After flying some of these simulators, I went to Warren North and told him that—well, I'll be nice to some people here, but basically I told him it was a bunch of crap. He said, "Well, what is your recommendation?" I said, "Well, McDonnell Douglas has got a good engineering simulator up there, and I know the people. One of the boys is a good K.U. graduate with me, so I know him, and some of the other guys I got along very well with." This is not to disparage anybody at JSC, but some of the engineers in the simulations group were not engineers per se that were interested in how orbital rendezvous occurred. They were interested in making their simulator work as a simulator, big difference.

I had tried to get the mission operations director division, and they had some good guys over there, and FOD [Flight Operations Division], that I got along with great. But their programs, I wanted them to run programs backwards. When we started rendezvousing from a different orbit and at a distance behind the Agena or any vehicle, you can do certain things presumably to try to do your rendezvous over a period of so many degrees.

A better way to do that is start from the vehicle, have them together and then run the thing backwards and find out where you end up with so that I can make a departure velocity and see where it put me here so that I could see what my corrections were in between on the orbit. They could not do that at JSC. But they could do it on the engineering simulator at McDAC at St. Louis [Missouri].

So I told Warren that I thought that somebody should go up there and run a bunch of these trajectories and find out what the variability was in terms of thrust intervals and increments of velocity inputs or subtracted to put you in the proper position at your correction points so that you ended up where you needed to be in order to rendezvous with the passive vehicle.

He asked who I suggested, and I kind of shrugged my shoulders. He says, "You're it," about like the LLRV/TV. So I went to MCDAC at St. Louis. Well, first of all Warren went to the program manager who was [Charles W.] Matthews at the time, told him what the problem was. Matthews agreed that if there was this problem, it would probably be a good idea if we put some more work on it. If Warren had confidence that I could come up with something meaningful that it was worthwhile to do that.

He was the Gemini Program manager. He called MCDAC and told them to turn their engineering simulator over to me to do what I wanted to do. So for the next three months I was at MCDAC in St. Louis. We did a number of these engineering runs, and as a result, a number of changes were made to the rendezvous concept. One was that the distance was changed. If you have a delta-H, which is the altitude between the Gemini orbit and the other orbit, if that delta-H is too small here, the trajectory that you have when you're trying to orbit with it, it doesn't take much to miss the upper orbit at all. You want a reasonable intercept angle to get there. If you have too big of an H, you can have a good intercept angle, but you can be either way ahead or way behind. So you have to pick the proper orbital transfer, it's called omega-T. That orbit distance is the orbit from this vehicle is here and you're down here in a different orbit, and you pick that delta-H and this transfer angle that you travel while you're going up in altitude to intercept the vehicle at a reasonable angle so that you can have a very small miss. The reason why you want a very small miss is because you didn't have very powerful thruster on the Gemini's own system, which would allow you to null out large errors.

So as a result of that we changed the delta-H, which was the altitude that we put the Gemini into trailing the Agena. The delta-H was the difference between the two orbits. We also changed the angle between the two from the time you started the initiation of the orbital transfer for rendezvous to the time you caught up with it. We changed that. They had some other things in there such as platform alignment, where halfway through the rendezvous you bring the vehicle down to local vertical to do a platform alignment for fifteen minutes and during that time you don't know where the hell the Gemini is and if you don't get a lock back on, radar wise, you're SOL.

So I eliminated that by proving to the powers that be that the error that we incurred by not aligning that platform during that thirty-minute period for rendezvous was negligible in terms of the amount of thrust it would take to null out that error and fuel that it would take. As a result of that change, we were able to look at the vehicle all the time, both with radar and with the acquisition light that we talked about earlier.

Now this acquisition light that we put on there was bright enough so that as they translated into darkness they could actually see that light blinking. We picked the intensity of that light such that we knew what the distance was and we picked the intensity so that we felt reasonably sure with a little dark adaptation after entering past the Earth's limb where it was dark and there's a little period of time there where you have a gray area. But after that point in time, they should be able to see that acquisition light. That was the whole intent of having it on there. Then, of course, as they approached closer, within a mile, then they could see the cone, that we adjusted the lights in the cone and then a little bit later then they could see the running lights that I put on the Agena to figure out what attitude it was in compared to them so they could null their velocities up and then come under it very slow and close without having to worry about big orbit adjustments.

During that period of time that we were up there, I had started this. Buzz came up and he was a big help along with a hindrance because he'd come up with some ideas that were just off the wall. It would take me, as I said, twenty-four hours to shoot down something that he thought up in five minutes, because I'm certainly not as sharp as Buzz in that area. Orbital mechanics was not my forte to start with.

But between the both of us, I came up with a set of manual procedures so that in case we lost the platform, that in case we lost the radar, that in case we had nothing but the stars to go by, I guess there were four different sets of procedures that I developed, where it would allow the crew to actually be able to rendezvous.

Of course the key to this was the automatic system was supposed to rendezvous by itself, with the crew putting the thrust in and reading the gauges. But no one had a lot of confidence in that, and there was no way to backup the system if it failed. So I developed all the backup procedures, and actually working with Tommy [W.] Holloway, who is now the ISS [International Space Station] manager, or whatever. He was responsible for developing flight manuals for that particular mission. Working with Tommy, we developed the actual flight manuals for the crew. While I was up there, we brought in two sets of crews. We brought in Schirra and [Thomas P.] Stafford and actually trained them on all the procedures so that if any one of these four things happened, that we could still make a successful rendezvous. We also brought up the backup crew which was Grissom and—[John W. Young]

Butler: I don't have that. I can look that up and we can put that in.

Grimm: I think it was Grissom and somebody else. Actually, for a short time we had [L. Gordon] Cooper and his crew [Pete Conrad] up there, this was Gemini IV, while I was doing this. This was like six months I went up there before the launch, and they [James A. McDivitt and Edward H. White II] were just getting ready to fly their flight.

It didn't surprise me that Gemini (?) you asked the question in here, was I surprised about Gemini IV (?). No, I wasn't surprised that they didn't accomplish their rendezvous, because they hadn't been trained and they didn't really understand the mechanics of rendezvous. It really wasn't their fault, but it's unfortunate that somebody hadn't picked up the gunny before, but that's kind of the way things were. Everybody was pressed to do certain things. This was one of the things, even though it was very, very important, not only to the Gemini Program but to our whole space program, it was falling down the crack.

Butler: You mentioned that Cooper and Conrad were able to come up and do a little training before their attempt with the pod (?).

Grimm: They didn't do any training. They came up and looked at what we were doing at the time, but theirs was in a different situation because they pitched this, as I recall, a ball out of the nose of the Gemini, or out of the back.

Butler: I think out of the back, yes.

Grimm: I believe out of the back. I'm not sure how far away it went. I think it was a mile or so, as I recall. Then they were not able to get back with it. They just flew around it. Of course, if you think about it, if you don't put your thrust vector in the right spot, the only thing you will do is just continue to make bigger circle until you run out because if you were nose to nose and you apply thrust to this vehicle, what you do is you increase the velocity of that vehicle. When you increase the velocity of the vehicle, it moves up.

So a lot of times when they were thrusting toward the vehicle, they should have been thrusting away from the vehicle, theoretically, which would have brought them down at a lower altitude and then they would have closed in on the vehicle. But it seems simple now, but it wasn't so simple then.

Butler: Nobody had done it before, and they were used to traditional flying in the atmosphere. Very different.

Grimm: So the crews, Wally and Tom, and the backup crews, both trained with the automatic system and with the backup procedures that I developed and with these changes in the techniques in terms of the omega-T the delta-H, the not aligning the platform. Again, here's where the COAS change came into effect, you know, that I talked about, the running lights, the docking lights. They kind of all work together in happenstance, I should say, to accomplish that. Astronauts responded very well to it. I had zero problems with it and they were enthused about it.

You say here, what technical problems did the rendezvous training present? It wasn't the training that presented any problems, because the guys were always there. I'd work with the engineering people after the crews left at 6:00 o'clock at night, and I'd be there until 4:00 in the morning. Then I'd go home and sleep for four or five hours and then be there about 8:00 to 8:30, and then they'd come in about 9:00. Then we'd start working with the procedures that I had developed overnight. That's the way we incrementally trained the crews.

The biggest challenge was that there was a gentleman at Headquarters who had guidance and control division. He found out what was going on and he raised such a stink with the Headquarters director that they came down to JSC and had a meeting with Matthews, and he and the Headquarters director of the Gemini Program. Warren and I came in from MCDAC and with Wally and Tom, Wally Schirra and Tom Stafford. They had this big meeting about whether my rendezvous procedures were any good or not. He said that my procedures had been done empirically, i.e., do one, if it didn't work, do another one and figure out what the difference was, correct that and so forth, and that we had to have a 100-percent guarantee that we were going to rendezvous.

We didn't have a 100-percent guarantee to start with, no matter what.

Butler: Not any of it.

Grimm: Not on anything. He was adamant about that. So before we'd come down, they sent Bill [Howard W.] Tindall [Jr.], who was the assistant director for ? I can't remember what the name of his division was at the time that he was in. It's like MOD or something close to that.

Butler: MPAD or Mission Planning [and Analysis Division]—

Grimm: MPAD, MPAD. I loved that guy. I'm sorry that he's gone.

Butler: Yes, we're very sorry about that.

Grimm: His Tindallgrams were just priceless.

Butler: If you could hold that thought real quick, I'll change out the tape. [Tape change]

Grimm: So they had sent Bill Tindall before they had this big meeting at MSC because he was over that group that I had tried to work with for developing the trajectories and procedures and so forth at JSC. They weren't able to, their computer would not do what I needed one, and they agreed with that. Bob [Robert W.] Becker was the guy that I worked with at that point in time, and [Edgar C.] Lineberry.

So Bill came up and sat with the crew and watched them for a day, and he said, "Yep, Dean, it works." So he went [home]. So when they asked Bill, oh, this guy at Headquarters says we need to get [The] Boeing [Company], who was the Headquarters contractor at that time supporting Headquarters, and they did theoretically all these analytical studies to give Headquarters the warm and bubbly feelings. He [the Headquarters guy] said, "We need to have Boeing take six months and run a good analysis of orbital rendezvous and Grimm's procedures and so forth, because we think it's snake oil." So he recommended that the orbital rendezvous thing be postponed, the mission be postponed.

Butler: This was already while the Gemini Program is in process now?

Grimm: Yes, we're like three months from flight on Gemini VI.

So he had this meeting and he made that recommendation and Gilruth asked Bill Tindall what he thought. Bill said, "I don't know how he did it, but the crew has never failed to rendezvous using those procedures." This guy said, "Well, I still recommend that we postpone the mission and do this analysis if we have to validate these procedures."

I was about at the end of my rope because I had worked a lot of hours and I'd spent a lot of time on this and you could call it snake oil if you wanted to, but they worked. I said, "Well, if that's the case, then I quit."

Wally said, "Well, if Grimm isn't on the program, I'm not flying the rendezvous."

Butler: That's nice to have that kind of support.

Grimm: So they kicked everybody out of the room except for the Headquarters guy and Matthews and Warren and the crew, and they came back and said, "We're going with the mission."

So I went ahead and finished up the training with the crews, went back to MCDAC and came back and developed the flight manual with Tommy and went over and sat in mission control and put on the headsets there and listened to what was going on in case Tom Stafford, who was doing the monitoring and telling Wally when to make corrections and so forth and to be ready in case there were any failures. So he was following right along the charts and Tom says it's tracking right down the line. Then they got the acquisition lights and then they got the docking lights and then they got the running lights and said, "We're here."

Wally said, "The damn things worked, Dean." Then they came back, they gave me something that they signed that I'll always treasure.

Butler: That must have been really nice for you, to see all of that hard work come together so successfully.

Grimm: It was, but it was stressful like a couple of these other jobs that I had always.

This was in '66, '65?

Butler: December '65 was the Schirra/Stafford mission.

Grimm: I'm glad you've got that. '65, and December '65.

See, during this time, I was also doing the project management on the LLRV/TV and crew station on the command module and supervising the teams, supporting the crews. I had teams assigned to every crew that was going to fly, and those teams were out at Rockwell or at Grumman. Wherever the crews were, that's where the teams were. These were about four or five different guys, and they actually participated in the checkout at the factory and replaced the crews when the crews weren't there. They'd sit in for the crews.

They were the ones responsible for working the crew equipment. They were the ones working the stowage, they did all the stowage allocation and keeping track of where everything was at. In general doing all those things that you don't want to have the crew doing because they are what I call grunt work, but necessary work. So we had teams assigned.

Then when I went on to the Cape [Kennedy Space Center, Florida], I had the same thing happen down there, and they were responsible for stowing all lockers, putting the lockers in, responsible for the suits, getting the crews fitted up, taking them out in the van, sticking them in the thing going up and kissing them goodbye. I don't know whoever that old German guy was who stuffed them in. I can't remember what his name was.

Butler: Guenter Went.

Grimm: Guenter, Guenter, yes. I never can remember his name, but yes.

So there was just a lot of things going on. My plate was certainly full during that period of time. I'm sure other peoples' were too, but the one thing that if I ever had to criticize Warren North about was that he didn't trust a lot of people, and you had to prove yourself to him. But once you proved yourself to him, there was no limit to what he would dump on you. [Laughter] And he dumped on me.

Butler: I guess proving yourself wasn't necessarily a good thing then, was it?

Grimm: That's probably true, but I was a glutton for punishment, and of course, like I'd said earlier, I always loved challenges. I was so into this program that I couldn't hardly wait to go to work every day.

Butler: That's nice. A lot of people can't say that.

Grimm: It was that way up until probably the last three years of my time with NASA.

Butler: That's really good. That's very fortunate.

Grimm: Of course, it's been that way at Boeing, when I was at Convair, when I was with FAA, I just loved my job and dug into it and said there's always things to learn and always to grow, always ways to do new things and there almost wasn't a challenge that I wouldn't accept. Fortunately I was in the position where you could do those things, and then I had a boss who had enough confidence in me to let me do it. That's a wonderful to have, because I would say that there's not many people that can say during their professional career that they had a boss who had confidence in them and let them do their thing without much guidance.

Of course, in many things we were doing there wasn't much guidance to be given because everything we were doing was new. It was a one-of-a-kind first-time project.

Butler: Never been done before.

Grimm: Never been done before. And thank God, we didn't make many mistakes.

Butler: It all came together so well, considering how little background there was in it all.

Grimm: So I think that covers the—there's some interesting things that happened during that orbital rendezvous training that I won't get into, but would make for a spicy book sometime. [Laughter]

Butler: Talking a little bit on rendezvous, later in the program, after it was proved out for the first couple missions in Gemini, then later they began testing different types of rendezvous, even one that was the direct rendezvous. Were a lot of those procedures based on what you did?

Grimm: Same basic procedures, except for the delta-H, you know, where we got brave, you know. In this direct rendezvous, for example, where the thing was going overhead and we launched and had to be right on the second and it came in. But the altitude it came in at was very similar to the delta-H that we had previously. Because the speed was different and the phasing different, the concept was the same but it wasn't quite the same in terms of the delta-H obviously or the transfer orbit in terms of degrees to travel before you actually rendezvoused. [Interruption, tape turned off.]

Butler: We're on.

Grimm: As far as problems, I think we've talked about maybe a political problem that we just talked about. From a technical standpoint, I think by the time we did a couple more missions, we finally had the simulator at JSC working right and then, of course, we were using the procedures that I developed for the backup procedures.

Actually on Gemini XII, the last Gemini mission, that's the one that's Aldrin was on, that they had a radar failure on that mission. They actually used the procedures on that mission to accomplish the rendezvous and, of course, it worked.

Now Neil [Armstrong], during his—he was already docked when he had his problem, when he had the runaway thrusters, the stuck thrusters [on Gemini VIII]. So we won't get into that. That hasn't got anything to do with rendezvous.

I think the biggest problem technically we had was understanding how a rendezvous really worked and then getting the delta-H right, getting the transfer orbits right, getting rid of some of the extra work in terms of platform alignments that we really didn't need and then with the proper aids in terms of lighting, the overhead light, the docking cone light, the running lights, the COAS, reconfiguration. Those were some of the technical issues.

Of course the crews were always great, and they always had a lot of ideas. Although they didn't have a lot of time to spend on solutions, they could tell you what the problem was. Of course then it was our responsibility to figure out a solution to the problem and then we'd work it out and go fly the simulation ourselves then to make sure it worked. Then we'd bring the crew in and they'd tweak it however they wanted some procedure or something or other. Then we'd put it in concrete and that would be it.

Butler: Did a lot of the procedures hold over as they were all proving out? Did you then transfer them over for Apollo? Were you involved in that?

Grimm: The concepts were very similar. The procedures changed because the one who actually started working the changes in the procedures and the detail of the procedure was Paul [C.] Kramer. He was the head of the mission planning branch in the division that I was in at the time under Warren North. He would work out those details. But the basic concepts, the crews now understood what they had to do, and instead of thrusting one way, they were supposed to thrust the other way and understanding the concepts of rendezvous.

We had a couple of astronauts which I was never sure that they really understood rendezvous too well, and that's about all I want to say on that.

Butler: That's all right. It is a challenging concept because it doesn't work the way you would think.

Grimm: No, because people tried to say we could brute force it, and you could have brute force and effect a rendezvous, [if] you had large enough thrusters to overcome the effect of increasing the thrust, if you knew which way to thrust to get the vector right, so that instead of going up you stayed on…course, which meant that you had to thrust [in a certain direction,] if you were going to add velocity in order to stay on the same course.

In hindsight, it's not a difficult concept, but at the time it wasn't understood well. As I said before, we didn't have any rendezvous procedures that worked, automatic or otherwise. At the time Langley [Research Center, Hampton, Virginia] had been working on rendezvous for several years, since they were a research center. They didn't have anything that worked, because I went up and flew their simulators. The Air Force had been working on it longer than that because they had the Blue Gemini, and later the Blue MOL [Manned Orbiting Laboratory] and they didn't have anything that worked.

Butler: It's interesting that it, with so many groups working on it that it hadn't come together more before this.

Grimm: I think part of the problem was that there wasn't the emphasis that it needed to be and I don't know whether at one time you might not have needed rendezvous if we'd have had Nova. There wouldn't have been one. But, again in hindsight, when we talk about staging and separation and the efficiency of the multiple stages and things creates a lot of extremely difficult technological solutions to be made. But in terms of being able to get off the pad and to do all the things that you need to do that that was the right thing to do at the time because our engines weren't that powerful, and they still aren't as far as that's concerned, except for the Russian engines which we've started proselyte off of.

So I think those were the major problems. By the time we got halfway through between Gemini VI and Gemini XII we worked out all the bugs in the simulators and in the procedures and in modifying the backup procedures and trying to do new methods such as the direct rendezvous for example. By that time I think we finally had a good handle on the concepts and what we needed to do for the various types of rendezvous that we wanted to do.

Butler: Where you were you? You mentioned with Schirra and Stafford that you were in the control room and following their mission that they were going for that first rendezvous. During the other missions, did you continue to go to the control room at times, or were you doing other things?

Grimm: No, I was doing other things at that point in time. I had people there in the control rooms but I was not there. I think that Holloway, Tommy or some of his people and Lineberry over in MPAD and Paul Cramer's group from the mission planning branc? that's not his title. I can't remember what it was? that was in our division, was the one working with Lineberry to develop the different rendezvous concepts at that point in time. I sort of passed it off because I told Warren this wasn't my area of expertise.

We proved our point and had been successful and had backup procedures and it was time for the people who had that responsibility to pick it up.

Butler: Of course, by the end of this you had sort of become somewhat of an expert in the area.

Grimm: More than I had ever envisioned, yes.

But so I think that kind of covers most of the things on the Gemini missions that I was involved in.

Butler: You mentioned that you were also involved with the crew stations, but that was primarily on Apollo; is that correct?

Grimm: Mainly on Apollo. As I said, the only thing I was involved with the crew station was the handling qualities for the docking area and the things I've already covered on the Gemini Program. I was much more involved in the crew station area in Apollo [Command Module and] in the Lunar Module area and so forth.

So I think you asked the question on the next line there about did I participate in actual missions? Only on that one, was I over there.

You asked what my thoughts were when they successfully achieved rendezvous. I had no doubt. I was very pleased, obviously, and I think I even got a commendation, I think on that. Should have been money, but it was only a commendation.

Butler: At least there was recognition for your efforts. Did that come from Headquarters? Were they finally recognizing that, yes, your procedures had—

Grimm: No, that didn't come from Headquarters. That came from JSC. I think Warren put me in for that commendation. At that time they weren't giving out the exceptional service medals, distinguished service medals like they did later in the program. In the initial parts of the programs, only the crews got it. A little later, only the wheels got it. By the end of the program, those things were being given out like stickers for the calendar.

Butler: You have a very good method of putting it all together, I guess.

Moving into Apollo then, and maybe we could talk some about you were involved with the subsystems on the command module and the crew stations. At what point did you become involved in this and how much of the design had already been settled on?

Grimm: I think I was a section chief for not over a year or so when the previous manager of the project support office left to go back into private industry. He wanted to be a Lear jet chief pilot so that's what he went to do for experimental flight testing, Bob [Robert F.] Berry. So that job was up for grabs. I had not been, as I said, a section chief very long, maybe a year. Even though there were other people who felt they were more qualified for that job, Warren selected me. Of course I always attribute that to the fact that I was the best choice. [Laughter]

Butler: I'm sure he would agree with that.

Grimm: He must have; he did it. In that job is where I picked up the responsibility for the command module crew station, the lunar module crew station, the team responsibilities, still flying the simulations, developing simulations, picked up the responsibility for the neutral buoyancy training facility, the air bearing training facility, all the mockup facilities and training in the mockups and all the stowage. Basically when I say all the command module subsystems, this includes all of the equipment, any stowage and the support of the crews and figuring out what goes where and when the crew needs it coming up with the GFE requirements, Government Furnished Equipment requirements and so forth.

This includes, of course, I told you about the teams where they would support the crews in terms of the spacecraft checkout at the factories and then down at the Cape. Then also we were responsible for stuffing all the stowage containers because we had made those removable after the fire and solid versus Raschel netting, which was one of the big problems in the fire. During this time I also had crews that man rated the large altitude chamber and the smaller altitude chamber over in SESL [Space Environment Simulation Laboratory], I forget what building [number] that is [Building 32], that we actually manned those capsules, walked out in a hard vacuum, climbed into the vehicle, shut the doors and then lived in there for seven days.

We actually man rated not only the capsule but the vacuum chamber as well for manned testing. The spacecraft that we had in there had a similar problem, but no fire, as the one they had at the Cape two weeks later [Apollo 1, AS-204]. We had a lot of water condensation in that vehicle. We had sparks in that vehicle from some of the wiring, but fortunately no fire.

Butler: [As] part of setting it up for the manned testing, were you also setting up some safety procedures in case of that type of thing?

Grimm: We went through a lot of pre-planning. We had an FRR [Flight Readiness Review], which I was involved in, in terms of looking at all the systems and the procedures. How we're going to get the crew in there, how are we going to get them out, which would have been very difficult if there had been a problem, just like in the actual spacecraft, because that was an inward-opening hatch in that spacecraft as well as the one at the Cape.

So yes, we were involved in all of that, including the team that I had in there, which was Neil—no, [Grimm addresses his wife] the guy that rolled you up in a carpet one night at a party.

Butler: Well, that sounds like an interesting story.

Grimm: It was a wild party.

Butler: At least everybody knew how to work hard and play hard.

Grimm: Yes, we did. But what was his name? There's Neil [Anderson] and Joel [M.] Rosenzsweig, who was an Air Force captain who died, and Don Garrett who left the program shortly thereafter. Joe [Joseph A. N.] Gagliano was an Air Force captain who was there just for a two-year period. I had a lot of those Air Force guys, sharp young first lieutenants, second lieutenants, came in and made captain and left.

Just as an aside, three of those gentlemen retired as four-star generals in the Air Force, and I met some of them here in the past few years. They recognized me, I didn't recognize them, but I was their boss at the time.

Back to the Command Modules here, until Apollo 7, I worked on the Command Modules. Well, I worked on them after that too, but in terms of design. …We had the fire about the time that I was on this program. We'd been working with the placement of all of the switches and all of the various instruments and the scaling on those things and the nomenclature on all of them, and guarding of switches was a big thing. Because of the crew floating around in there up against the instrument panel and so forth, you could very easily throw the wrong switch, which happened occasionally. Figuring out where to store things and how to store them, and of course after the fire we went through a big redesign, I forget what that redesign cost us, but I think it was like a hundred million dollars, as I recall. I don't know whether that's a good number or not, but that's the number I recall.

I'd go out every Monday morning and come back every Friday evening, and I did that for a year, to Rockwell. This was at the time when Frank Borman was the lead out there initially in terms of the design. He had the responsibility for directing Rockwell on all the design changes that we were going to make to the [Command] Module. Later George [M.] Low, who was at Headquarters, came down to be reassigned as the Apollo Program Manager when Joe [Joseph F.] Shea was moved to Headquarters after the fire.

So we worked daily with the Rockwell people out there in defining closeout panels, defining stowage, stowage lockers, modifying the seats, working with GFE in terms of changing all of the flammable materials during fire tests out at White Sands [Test Facility, New Mexico], changing the flight data material so that the paper wouldn't burn, changing the suits so that they wouldn't burn, getting rid of all the nylon or the Raschel netting that we had had in there before, adding, changing. We did a lot of system redesign as well as that outward-opening hatch. System design and software redesign, more testing on a lot of the various systems, subsystems. As I said the placarding of instruments, the gauges, guarding various things with flip-over guards for switches and other things to keep crews from snagging on things.

So there was a lot of work that was done after the fire during that period of time and up through the actual mission. So I think that covers the subsystems that I was involved in, all of the instruments, the scaling, all of that normal stuff that's associated with all the instrumentation. We were involved in the telemetry instrumentation, instrumenting the various things that we wanted more instrumentation on and so forth.

The spacecraft was gone through from end to end basically in terms of handling qualities, characteristics, software, backup systems, crew equipment, storage, safety items, outward-opening hatch and so forth.

After that was over, I brought in another gentleman to take over that responsibility because I was getting loaded up again with stuff, still [had] the LLTV/RV was on my palate, a big load. His name was Chris [D.] Perner, very exceptional, laid-back West Texas gentleman who was a super, super good guy. He became about three iterations later a division chief of the division I had before I went up to the director office.

That's a long story too, because I hadn't gotten a division yet. Then I moved from flight crew ops over to the engineering directorate and had a division. Then I had a guy follow me who was my assistant chief, who was Don [P. Donald] Gerke, who died of a heart attack about five years ago when he was at Headquarters. The next guy in there was George [C.] Franklin, who was my subsystems manager on the lunar module. He did all the same things on the lunar module that I was doing on the command module at the time. I don't know if you've ever interviewed him or not, but he would be an interesting guy to interview.

Butler: Not yet.

Grimm: Both Chris and George. They were both subsystem managers under me, very capable, competent people.

Butler: It's fortunate that you were able to have people like that to work with.

Grimm: Yes. It's nice to be able to select people—you couldn't do that now, but you could back then, the people that you'd like to have work on these types of assignments. So I think that covered that.

The most difficult part, the question here you have, of working on these subsystems was since this was a crew station, one of the most difficult things was working with the crews. The crews had definite ideas about things and for the most part they were able to get their way, although we did have some very interesting discussions about how to implement certain things in the crew station.

The other thing was that we were on a tight schedule during this rework after the fire. Money was always an issue and the changes, there was always an "Is this mandatory?" Is it a safety issue, safety flight issue, or is it a crew issue or is it a system issue and you had to define what it was and what the impact was if you didn't get the change. Then you had to go before a change board individually each one of us and justify to the program manager who at that time—now is George Low—and justify these changes. Earlier in the program Borman was making changes before George Low came down from Headquarters where he was the associate administrator assigned to this job.

Obviously you never got to interview George Low or maybe Ertel interviewed him, but he was a very exceptional individual. Young guy, he wasn't much older than I was. He emigrated with his parents from Austria. I was trying to think of the school he went to, [Grimm addresses his wife] remember I was telling you the other day. It was Rensselaer [Polytechnic Institute] that he went to. A very sharp individual. Then he went to work for—what's the center at Cleveland? It's Glenn now, but it was Lewis [Research Center] then. He was so sharp he got selected early in the program when NASA became from the old NACA [National Advisory Committee for Aeronautics] under [Robert C.] Seamans [Jr.] and went up there as an associate administrator, and then deputy administrator. Then after the fire they wanted somebody with a good technical command of the situation as well as administrative management to come out and manage that. He actually lived out at, had a cot in an office out at Rockwell, and lived there twenty-four hours a day for the better part of a year.

He was out on the floor at night and during the day and he looked into everything. He knew more about that spacecraft in total than any of the rest of the people in the program by the time we were ready to fly again. As I said, a very exceptional young man.

Butler: That's one thing with the oral histories, that even though we haven't been able to talk to him, we have been able to learn a lot about him and his contributions.

Grimm: So I think that kind of covers that.

While we are talking about Apollo, do you want to jump to the fire? I think you had a question here.

Butler: I know you've talked a little bit about some of the redesign here and obviously that was a difficult time.

Grimm: Right. After that. [Interruption-tape turned off]

Butler: Okay, we're going.

Grimm: Apollo 204, and I didn't remember that being the name, was the fire down at the Cape. What had happened there after that fire, was there were a lot of people concerned about their political careers and their butts and Rockwell and some NASA folks in particular. Before the accident committee board got totally formed, Gilruth sent Borman down to be the representative from JSC even though this was the KSC [Kennedy Space Center]. KSC basically has responsibility for the launch but JSC was still responsible for the vehicle and the crews.

So the board was formed under, I think, [Floyd L. "Tommy" Thompson] who was the director of Langley at the time. Borman was the technical interface—who ended up being the technical interface with the board because most of the people on the board were external to NASA or at least external to the space program. So they wanted somebody who was technically knowledgeable and Borman was a good systems person.

He always had an authoritative manner about him, so I guess they figured he was a good one to be down there. He started having a lot of problems with various organizations trying to maybe slant the direction that the investigation was going and also to find out what was going on, even though they had no legal business to find out what was going on at the time that the board was working.

So as they started to go to grid in the spacecraft and get ready to move things out, there were even more problems. Borman said that he needed somebody down at the Cape that he could trust. So Deke and Gilruth had a short conversation, and Deke called me up and said, "You're it." Of all the people that he could have chosen, there were a number of other people

I suppose he could have. But they decided to choose me and so it was my responsibility to go to the Cape and take over a hangar down there and then [have] responsibility for [the CM] after they had put the grid in the spacecraft to photograph everything, and then to also use this metal framework to remove the crews in the seats and then to get to everything else. You can't get in that spacecraft in a vertical position without standing on everything.

So we put in this metal grid that was built very quickly to allow them to remove the crew members and then to start removing things piece by piece. It was my responsibility to take possession of everything and to lay everything out in a grid format so that they could locate everything in the spacecraft in the hangar, so everything would be relative to everything else three-dimensionally and be able to have access to it in case they needed to have work done on it or to take items out someplace for what I call forensic inspection and examination.

So that was why I was down there, and that's why I was chosen, as best I can tell you. As far as my previous background with the FAA in terms of actual investigation, I don't know that that had anything to do with that. It was more my knowledge of the command module and the subsystems and what was everywhere plus the fact that I think that they felt that I was the one [or] at least one of the people that they could choose who could not be compromised. There were attempts at that while I was down there.

Butler: That's unfortunate.

Grimm: That gets back to some other things later on here, where you said why did I leave NASA, and I'm not going to go that direction because it has to do with some of the people that are still at NASA.

Butler: Sure.

Grimm: So I think that pretty much covers it. I was down there for almost three months, I think.

Butler: Must have been a tough job, although it must have been good to know that they could rely on you. That they did rely on you in that way, it must have been hard to get through.

Grimm: Well, it was. Of course, I knew the crews quite well, since I've worked with them for probably three or four years with Gus and on the Gemini things and the training and the backup training on the rendezvous. Of course I saw them on a daily basis because I was in the same building with them and I processed all the changes that they wanted since I was the subsystem manager on the Command Module and the Lunar Module. I signed off on all the proposed changes, and I went and pitched those changes to the Change Board to Low, and I always put out a weekly letter to all the astronauts letting them know what was going on in terms of activities related to the spacecraft changes, getting their inputs on changes and trying to take a total overview of the crew organization to the change board and presenting it to the change board, to Low's board at the time.

You asked what was the most difficult part of the job for me. Well, it was trying to make sure that nothing within the building got compromised, physically or otherwise, and making sure that I kept Borman informed of what was going on. Of course, I assisted the accident board members who wanted to have a technical tour of all of the parts of the spacecraft and so forth. Then I'd write up daily reports of any sequence of events and what came in from the spacecraft and so forth so the board knew what was going on through Borman. It all went to Borman and then to the board.

So that was most of the activity associated with that job. It's hard to believe, but we had some Headquarters people come through there and some of them wanted souvenirs. I had to practically break some knuckles to keep their hands off of things. During that period of time I was offered a few bribes by various organizations, as well, which that's as far as I'm going to take that subject too.

Butler: Sure, sure.

Grimm: Does that cover what went on down there?

Butler: I think that covers it pretty well.

Grimm: Let's see. From there we jump to, this is kind of backtracking, because we went to the fire and out at [the North American Rockwell plant at] Downey [California] for the factory for that year after that. Then I got, as I said, Chris Perner on board and [George] Franklin was very capably managing the lunar module stuff at that point. I had some good people on the trainers and mockups and the water tanks, and some other people that were working the man rating and manning the spacecraft and the altitude chambers.

I was working the LLRV/TV also during this period of time.

Butler: You mentioned a couple of the different systems and the mockups and the simulators and the trainers. You mentioned that the water tank, the training there. Were you also involved with some of the testing like where they would have the astronauts in the suits and try and simulate the one-sixth gravity for that?

Grimm: Oh, yes, oh, yes. That was what the water tank, what we did at the Boeing facility, put mockups in the tanks, and then all of the activity was in my organization.

Then we had a group in our organization that was actually a shop, and they had a very exceptional person who could build anything. He built that for me [Grimm gestures to a space shuttle model] when I left the organization. But he could build a mockup with all the functionality that you needed to do the training that was required. He had about twenty-four people working in the woodworking and machine shop. That we made everything there to support all of our training activities with the mockup and the neutral buoyancy facilities, because the neutral buoyancy facilities had a different function than the mockups that we had in the 1G in the 1G trainer, in that it had to be floated, had to have flotation and a lot of the functionality that you'd use for storage maybe or some other function that you didn't have to use or couldn't use because when you put the crews in the suits, of course, they had to be weighted in all three axis so that they didn't spin like a top or something or other. They had to be weighted with weights around their waists, around their arms, around their ankles to give them not only neutral buoyancy, but they also had to be able to be vertical in the water rather some other angle.

Then of course we had divers in the water all the time because we had to make sure that everything was safe too. We monitored their work and then if they needed assistance we'd work it. Because in the tank as well as the ones in the trainer, we were always working procedures, working bugs out of the equipment, working bugs out of suits, out of mockup cameras, handling all of these things that you have to work with while you're in that sort of activity.

I think about somewhere in this period of time, the organization was split up. I was going to say split up, but it wasn't at this point in time. This was still back in '67, '68, '69.

So, as I said, I had teams of people that were supporting the crews at the factory and down at the pad. I had teams of people supporting the crews while they were doing their neutral buoyancy training. I had another crew of people that were supporting the 1G trainers. I had, of course, Chris Perner and George Franklin supporting the crew stations in both vehicles, another set of crews, two sets of crews that were doing the manning of the [altitude chambers]?we did it twice, as a matter of fact.

First of all, we man rated the altitude chamber and the [006] spacecraft in a chamber? Then the LM, I forget what the number of the LM was that we had in there. LT[A]-8 sounds good anyhow. You might want to check that.

Butler: We can check that one in the papers, too.

Grimm: LT-8, and we manned that on at least two different occasions, checking out various aspects of the procedures, suits and systems. Then at the same time, I had another group of people that were working experiments and a number of us were still flying simulations. So that was kind of a smorgasbord of activities that I was doing while I had that organization. That was the Operation Support Office under Warren [North].

Butler: It must have been a unique challenge for you in a way to balance the engineering, which had been your training and your background with all this management, and to make the two mesh.

Grimm: I never considered it a problem. A lot of people that worked for me, worked with me, some of them to this day would accuse me of micromanaging the organization because I looked into everything. But by the same token, what I told some of these people was that when I can see that you're doing your job, you will no longer see me. But if you manage your job like you're supposed to, then you won't have to worry about me micromanaging your part of the organization.

Butler: That makes sense.

Grimm: That's exactly what happened. I said, "As long as I see problems with your organization, you're going to see me looking over your shoulder. When I no longer see problems, you won't see me." That's kind of the way I operated, although I had a very good knowledge of what was going on in all of these areas, mainly because I was interested. Some people do it from a management standpoint and never get their hands dirty. I was out there, I'd be walking through the organization at midnight.

I've got a hard hat up there that they painted red, and it says "The Chief" on it. The reason why they had that hard hat painted red is because they were surprised so many times by me, unfortunately in many cases, that they gave me that hard hat. I would wear it so that they could see me coming from a distance. [Laughter] So I've kept that hard hat up there.

Butler: That's pretty good.

Grimm: So I guess then for the next couple of years, even though all these other things were going on, I would spend a couple hours in the office every morning and have a staff meeting every morning like at 7:00 o'clock, get a status of what was going on, and find out what things I had to be doing, and then also be involved in the LLTV project at that time because we were getting the [LLTV] vehicles in, we were trying to get the facilities up, we had Neil's [Armstrong] accident.

The LLRV program had been under [Joseph S.] Algranti under an engineer that he had. As a result of that accident, they decided they wanted to change management, so they put everything under me again, both the RV and the TV. So we flew the RV for a while and because there was so much—well, another long story.

They wanted to, somebody had decided, I think mainly the crews as I recall, didn't want to be traveling back and forth from MSC to the Cape as they approached the flight, which is a good philosophy. So they wanted a facility down there with an LLTV down there. So I started working with the Air Force people, because this would have to be on the Air Force side by the skid strip and went in off of the skid strip. So I started putting together a management plan and working with the facilities down there and our facilities people to put in the requisite hard facilities there such as the H_2O_2 tanks and hangar and all the fixed facilities that you have to support a vehicle like this, that's this sophisticated.

We had pyros and we had an injection seat, which was some more pyros, we had JP-4 fueling, we had oxygen on the vehicle. We had H_2O_2, 92-percent pure, which is very explosive if it comes in contact with any contaminant. You drop an eye dropper, one drop into a bucket with dirt in it and it explodes. So it has to be very careful the way you handle it. You have to have a lot of cleanliness, all of your tubing and everything has to be super clean. You have to have people in suits with breathing apparatus and so forth because of the toxicity. There's a lot of requirements that you have associated with a vehicle like this that people are probably not appreciative of. A lot of people thought, "Dean, just go build this vehicle, build the flexibility in. When we get the LM, we'll tell you what the requirements are. You tune in the requirements, the crews will come out, kick the tires, jump in this sucker, and learn how to make that transition and pitch-up maneuver and come to a hover and land. Nothing to it. Piece of cake."

Butler: Not quite that easy.

Grimm: Not quite that easy, that's right. We were limited on budget, and the schedule was way too short to do what we did. So I spent a lot of time, as I said, after Neil's accident, he said, "What happened?" "Well, Neil, you ran out of nitrogen there." One of the things that we found out is that when you fill those tanks, you pump nitrogen into those tanks, those tanks get very hot because you're pumping this gas in through an orifice. You find out that you don't get the tanks full.

One of the things we learned is that we put wet towels on there and kept the tanks cool the whole time we were filling them. We filled them at a lot slower rate and then you let them cool down and then you'd fill them some more, then you'd let them cool and you'd fill them some more. So you ended up with full nitrogen tanks. Those are the two little things that look like two little male elements at the back end of the vehicle. Those are the gaseous pressurant that we used to pressurize the large H_2O_2 tanks that are the sides of the vehicle, those big large tanks. That nitrogen gas forces that H_2O_2 out through the lines and to the thrusters. Thruster's very interesting, it's a very simple device. It's got a platinum catalyst bed inside of that little thruster. The thruster is adjustable from twenty pounds to ninety pounds, and when this H_2O_2 hits this platinum catalyst bed it decomposes into steam and water. You get your thrust basically out of steam. That's what, if you've ever seen movies of the LLTV, this big clouds is this hot steam coming off this catalyst bed and coming through this little orifice nozzle, which provides the thrust.

So once it gets to that point, it's very safe. It looks scary, but it's a very safe compound. It's a lot safer than hydrazine or some other bipropellant material that you'd want to use to be around crews, which is toxic. I mean some of the other material is toxic.

So what I determined shortly after that accident was that we were trying to fly the RV and trying to put together the TVs. We were limited on resources, we were limited on facilities, and at the same time I told Deke after exploring all of the considerations down at the Cape on the skid strip that there were too many things that would restrict our operation down there. The Air Force being the first one, because of their safety requirements and their operational requirements. Then Air Force and NASA using the skid strip to land on, you had all kinds of constraints there. It meant more facilities, more people, splitting the technical people, splitting the operational people so that we'd have to be training in more places. I said, "We just don't have the total capability to do it and we don't have the time."

I recommended that we cancel the activity down there and Slayton and Gilruth agreed. So we did that. The next decision I made is that I can't fly the RV and TV, operate one and be building the other three at the same time, because we had two RVs and I'd purchased three TVs with this completely difference design that I talked you about.

So I made the decision that we were just going to ground the RVs and go with the TVs. That caused a little delay, but at least we were then able to put all of our emphasis and our manpower into the TV. To me it made sense at least, and to both of the other people that I talked to that I had to convince because the TV had the characteristics that we needed built into it or adjustable so that we could those characteristics as a LM into it so that we could train the crews.

The RV would get us a small sampling of what they needed, but would not give them an exact thing that I felt we'd designed the TVs to do. So as a result of that, we grounded the other RV and proceeded to put together two LLTVs, which we did, and which we started flying. We had quite a number of test flights on the TV. The day before we were going to have—the same day.

The same day Joe [Algranti] was going to fly this vehicle and we had the second vehicle ready to fly, and Neil [Armstrong] was going to fly that vehicle. Well, that's the day that we had the accident. So obviously, we didn't fly the second vehicle.

You say, what happened. Well, what happened was that we had a wind restriction on the vehicle and we instruments on the vehicle, Pitot tubes and anemometers. But between the ground and the altitude we had a wind shift and it wasn't noticed at the time, apparently not noticed. So as Joe accelerated to the velocity he was supposed to go into to go into lunar sim mode and to do the lunar descent simulation, he encountered a wind that was not a tailwind, it was more like a headwind.

So we exceeded the moment capability of thrust out of the thrusters to control the attitude. In other words, the thrusters were firing a hundred percent down, and the wind was pushing hard enough on the vehicle that the force exerted by the wind was more than the force of the thrusters to control the attitude of the vehicle. So the vehicle pitched up and rolled, ninety degrees on its right, and Joe was trying to correct the lift and unfortunately that command continues to build as long as you hold the stick in. Well, eventually the vehicle got control of itself, but now all that input that had been made came over so hard that it rolled 180 degrees. It went from ninety degrees right to ninety degrees left.

By that time he was putting in the other control to the right. Eventually it came back, but by that time the vehicle had built up a good descent rate and the vehicle was totally in control, except for the descent rate, when it hit the ground. The control system had done what it was supposed to do to stabilize the vehicle. Basically Algranti ejected out of the fireball. He said, when we took the film and analyzed it, it was around six-tenths of a second before the vehicle hit the ground. But the vehicle was coming down at approximately 160 feet per second. So he actually ejected before it hit, but when I was standing there looking at it, it looks like the thing hit the ground and fireball and he came out of it, but in fact he was coming out of it as it hit the ground. But your eye just couldn't pick it up that quick, but the film did.

So when you say how did this accident differ from the first one, I think I've explained the differences already.

Butler: Sure.

Grimm: Wally Schirra was the local board chairman, and we had Headquarters people sitting in on it and then people from Headquarters second-guessing it. There was talk about Headquarters canceling the program. Crews, Neil and Pete and Wally, who hadn't flown them but was the board chairman, Gilruth, North, Slayton, there was a number of people involved, including myself about do we have an inherent design problem in the vehicle.

Well, the answer was no, we didn't have an inherent design problem, but we did have an operational problem and we had to impose more stringent requirements on how we flew the vehicle because we knew the limitations of the vehicle. Or I certainly did. I think we transmitted that information to the crews. So as a result, we made sure by getting an Air Force theodolite [and balloon] team out and they sent up balloons half an hour before flight and I think we did it at five minutes before the flights. I think we did it half and hour, fifteen minutes and five minutes before the flight. So that we knew what all the winds were aloft to a thousand feet. Then, of course, we did one right after the flight, because believe it or not, we had a lot of wind changes in that thousand feet in a period of five or ten minutes because we flew early in the morning when everything was calm, no thermals, and hopefully no thermals and so wind sheers. However, we did have inversions and we did have wind sheers within that thousand feet occasionally.

So that the ground was saying that the wind is in this direction and at a thousand feet it was in another direction. So after that, besides the satellite [meteorology] team…[the] Air Force had a [helicopter] rescue crew out there every day to do a crew rescue in case the vehicle crashed, as they were there when Joe ejected, and they [also] used the vehicle for fire suppression due to the [rotor] blast, plus they had fire suppression capability onboard the helicopters, [as] well, in terms of foam and so forth and rescue people in the rescue helicopter. Then, of course, we had two fire trucks on the ground.

Then we took the third vehicle, which was basically the frame, we're now down to two vehicles. We took the third vehicle and took it to Langley and put it in the wind tunnel and validated what I think we already knew, which was the fact that if you yawed this vehicle to the right with the hole on the [left], and the right-hand side of the vehicle closed in, that that just acted as wind vane which tended to turn the vehicle. Well, based on our jet logic that we had put into this vehicle to simulate the lunar module, we did not have enough thrust to counteract that torque or that moment.

So the design that I'd had, which was to put a top on this thing and close in the sides, including the front, was probably not the best idea that I ever had. So as a result, we cut a large hole in the top to let the air out and to prevent this large pitch and yawing moment. The pitch wasn't so bad, but the thrusters had to handle the pitch and yaw at the same time and it made it very difficult for the thrusters because we had the thrusters set down at a lower limit to simulate the torque to inertia ratio that the thrusters had that were on the lunar module. Even though we had more capability, we didn't have the system set up to use it because we were trying to simulate the LM.

We had a backup set of rockets on there, but unless you went hard over on the hand controller, that didn't come into play. Of course, this happened very quickly, this upset condition, and that's what caused the problem.

So from then on we measured the wind incrementally. We put extreme strict limits on how much wind there could be for first flight, second flight, third flight, fourth flight, how much wind we could have at altitude, what the forward velocity could be max, if there was any crosswind, and modified the cockpit to take most of the top out and did not use the front Styrofoam piece that simulated the window view. Took that out. We did a number of test flights, checked out the system with Algranti and Bud [Harold E.] Ream and said we are ready to fly.

After about a month going around and around the horn, Gilruth got Headquarters to agree with that. Headquarters was all for canceling the program. They actually told Gilruth, "If you fly this vehicle, it's at you own risk." Gilruth essentially sent a TWX [teletype transmittal (pronounced twix)] back to Headquarters saying, "Tell us something we don't already know," which said he had confidence in the fact that we knew what we were doing, as much as anybody could know what you're doing with a vehicle like this.

We did the test flights with Algranti and Ream, and now we're about a month prior to flight, the lunar landing. So I called Neil at the Cape and he was down there doing spacecraft checkout. Neil Armstrong, we're talking about. I said, "Neil, I think we got all the bugs worked out." He wanted to fly the vehicle. He had told program management that it was a requirement for him to get so many landings in the LLTV prior to the go-ahead for the lunar landing. He felt that strong about it. He had flown the RV quite a bit, and had flown the TV, as had a number of the other astronauts.

As a result of the decision to fly and my calling Neil, I told him we needed to have him up there on a certain day. I can't remember what day it was now. He thought he could work it in his schedule. I said, "If God willing and the creeks don't rise, maybe we can get you enough flights in over this weekend and the next three or four days, if you can spend that much time with us, to give you the confidence that you feel you need to tell the FRR board, which was just getting ready to convene, that you're ready to go as far as your training is concerned."

So he flew up that Friday, and he was under a lot of stress obviously, from a lot of different sources. Of course, I was too. I think everybody on the program was. But I was getting calls from Senator [Clinton P.] Anderson's aides, and I was getting calls from Representative [Olin E.] Teague's aides, each one of those were head of committee in their respective branches of the government for NASA.

I was getting calls from program management at Headquarters every day. If something happened on the vehicle, quality would tell their boss who was another director, that director would call Gilruth and Gilruth would know about it before I did. This would be in the middle of the night. So I would have to, if I got home at 10:00 o'clock, I'd be back out there sometimes at 11:00 or midnight or 2:00 o'clock in the morning and be there for the rest of the day.

Be that as it may, we overcame all those little disturbances in the program and Neil came up. The best I recall, we had never gotten more than two flights in a day because of either a vehicle problem with the checkout or with the vehicle or with the wind, weather. But the first day he got there, they got in three flights. The next day, we got him four flights. Then on top of that—I don't know who did it. I still think Gilruth did it, but I can't at this point in my life remember for sure who did it. They set up a press conference out there. So we had all of those people in our hair in addition to everything else we were trying to do, which was get Neil trained.

Well, anyhow, we pulled it off. He got his flights, he came down, he said, "I think I feel comfortable now, having these many flights this close together." We'd moved him up in terms of different things to look for to get him closer and closer and closer to the lunar descent and the velocities and so forth. So we had the press there and Neil got right off the vehicle and went and talked to them, as if it was a matter of fact thing. It wasn't matter of fact for any of us, including him. He didn't like to talk to the press anyhow. So the press left then and we had our little debriefing with Neil and he left.

I told everybody to wrap up the vehicle because it was going down for a couple of weeks. I had stretched the time that we were supposed to do a certain number of checks on this vehicle periodically, just like you do on an aircraft. I had postponed some of that stuff that theoretically was mandatory. So I told them wrap it up, stick it in the hangar, we wouldn't fly for two more weeks at a minimum. We had some mods that we had to make that I'd postponed and things that we were just getting by the skin of our teeth on the vehicle.

Then I went back and sat in the office. Everybody else was going out and having a big beer bust. They asked me if I was coming, and I said, "No, I'm not coming. I'm going to unwind," and I had something else to do anyhow. I was waiting for a call from Headquarters, and it came about an hour later. It was General [Samuel C.] Phillips, who's the director of the Apollo Program. He said, "Dean, I've had a patch put through from the airplane to the ham operator at Andrews [AFB, Maryland] and he's calling [you] on landline and I want to know what happened today because we're on our way to the Cape, all of us and the management here, on our way to the Cape in G1, [and] G2, whatever it was." G1, I think it was, they call it the Grumman airplane. "For the FRR tomorrow and make a decision on whether we're going to go for a lunar landing here next week." And he says, "We've got to know how Neil's doing on this training."

I said, "Well, General, he had three flights yesterday and four today." That's the best I can remember in terms of flights. "He had his last one today at whatever time it was and he's satisfied that he's got enough training and he's ready to go as far as this vehicle is concerned." General Phillips said to me, he said, "Well, Dean, if you could keep—" I probably exaggerated the term, but I said "bucket of bolts," but I think he probably said, "If you can keep that thing together long enough to give Neil seven flights and he feels comfortable with it, then we probably won't have any trouble making a lunar landing."

Butler: That's nice.

Grimm: Then, of course as I said earlier, Neil came back as Pete [Conrad] did similarly [and said] that they felt the LLTV training was essential in their having a good feel in what was happening and how to handle the view out the window and the pitch and the control of vehicle to null out the velocities for landing.

Butler: An important part of their training.

Grimm: Of course, you said was there talk about canceling the program. Obviously there was. As I mentioned, MSC people were supporting the program, obviously they did. Gilruth was a little antsy about it, but the crews and Deke said that they thought it was mandatory. Of course, Neil was one of the foremost sponsors of that. They knew it was risky, but everything in the program was risky. So they were willing to take that risk. Of course we had proved that we had a good ejection seat on the system.

Butler: It certainly did work.

Grimm: It was supposedly only qualified for 120 feet per second vertical descent, and I'm sure Joe's was about 160. So we know that it worked quite well.

Everything on the vehicle was a modification of something. Some of it was commercial hardware. Very little of it was mil spec [military specifications]. Some of the stuff I wouldn't even want to tell you where we got it [from] to put it together, because it was such a unique vehicle that we just got it from every place. The engine was modified, wasn't made to run vertically. We modified it to run vertically. The ejection seat that was modified, weighed initially 150 pounds, we knocked off a hundred pounds practically and put on a different rocket off the rails. We did just so many things in terms of just the instruments and controls and displays and the avionics that had basically three analog computer systems operating the vehicle.

So to my knowledge, it's the first "successful," in quotes, fly by wire, nonaerodynamic vehicle that's ever been built. There had been other vehicles, and we lost three vehicles, but we never lost any crew and more to the point, no astronauts, which would have been very devastating to the program, had we lost anybody, even more so if they had been astronauts.

We ended up with two TVs after Joe's accident, and then we lost another one with a, believe it or not, product improvement on the vehicle.

Butler: Before you tell me about that, if I could change the tape real quick?

Grimm: Sure. [Tape change]

Butler: Tell me about the LLTV incident, which actually was a result of changes that were being made.

Grimm: As I think I mentioned on the previous pause here, the cause of the next accident was as a result of a product improvement by the manufacturer of our alternator/starter, which was a combination device on the LLTV. This particular new alternator, I can't recall all the details on it, but I think it had more capability in terms of power.

But one thing that we didn't know was that if the alternator failed, we had batteries on the vehicle at backup to power with enough electricity to get us down on the ground. Of course we didn't have much flight time, anyhow. Our total flight time was like seven minutes on this vehicle, of which only two minutes were in lunar sim mode to landing.

To make a long story short, this alternator had residual voltage in it when it presumably wasn't operating, so the residual voltage in the system was enough to keep the backup system from switching in and going on batteries. So it didn't let the system switch to go on battery, so that the electronics would work and therefore the pilot flying the TV, which was Stu [Stuart M.] Present, had no control. So we lost the vehicle because of that, quote, product improvement. It's one of many times in the history of technical things in the United States, where product improvements have come back to bite us in the ass.

A lot of times if you have something that you have problems with but it works, you're better off to keep it unless it's so bad that it's a safety of flight type thing. Our other system was not a safety of flight. We could have very easily have continued to use that system and not changed it out. But I was off the program at the time, and I don't know what decision I would have made had I been on the program, because once Neil is down and I had told the ground crew and the other program manager who was coming in to follow me what needed to be done on the vehicle, and I had signed off everything over to him, which was Charlie [Charles R.] Haines, in '69, I told Deke that I had had it. I had been on that program as long as I could stand it, I had done what he wanted me to do. I was totally stressed out, and I wanted off of it. I got off of it.

I took a month's leave and when Neil landed on the Moon I was sitting in Colorado Springs [Colorado] in my brother-in-law's apartment. I was a three-pack-a-day guy those last two years. I had made a promise myself that when Neil landed safely on the Moon, I would quit smoking. At the instant he touched down, I stubbed out my cigarette and have never smoked again.

Butler: That's great. That's wonderful.

Grimm: That's just a little aside. It doesn't need to get in this.

So I needed that month break. I told Deke I was taking off on a month and I didn't care what happened. I needed a break because I'd been on that program basically as I said, twenty-four hours a day, seven days a week, with no leave, no nothing. My children didn't even know where I was at because I was coming home after they went to bed and I was long gone before they got up, so I didn't see them for months at a time. They'd ask my wife whether I was on a trip or whether I was at NASA. But that's the way it was for a lot of people. I wasn't the only one in that position.

So anyhow, after I got off the LLTV program, it obviously continued for another year and a half with one vehicle. It was very successful and they concluded that program. The Smithsonian thought so little of that vehicle that they said they didn't want it when it was offered to them. They still don't have it. They had it up there for some anniversary, and then it was that Marshall [Spaceflight Center, Huntsville, Alabama] wanted it. JSC let Marshall have it for a couple of years, and then it went to Headquarters for some anniversary, whatever that was. Then, finally, JSC's got it back, I understand. The RV that was flown very few times is back out at Dryden, FRC as I called it.

So I think that concludes it unless you have any other questions on the RV/TV.

Butler: I think that covers it pretty well.

Grimm: May be more than anything you ever wanted to know.

Butler: Oh, no. I've certainly learned a lot from it because there isn't a lot of good information down about both programs and about some of the intricacies behind them and the challenges. It's been very informative.

Grimm: From that point on, I had started already before I left the TV program, started working some aspects of the Skylab program. At that time, JSC didn't have any trainers and we had simulators but no trainers. Skylab was such an interesting mishmash of modules, none of which lent themselves very well to a simulator like the command module simulator, the AMS, the Apollo Mission Simulator, or the Shuttle simulator or things like that because other than a control system it had very little in the way of very sophisticated controls in it.

So I had a number of good arguments and discussions, maybe is a better word, with Reg [Reginald M.] Machell, who was the head of an organization within the Skylab program office, and at that time Kenny [Kenneth S.] Kleinknecht had transitioned from the Gemini Program to the Skylab program as a program manager at JSC. I had a number of conversations with Kenny.

At that time we were having a real dogfight with Marshall, because Marshall had built a water tank down there that was an out-and-outright misuse of money allocated by Headquarters. Within this hangar they had built this forty-foot deep by, I believe, 100-foot-diameter, or eighty-foot, maybe it's a hundred, whatever, water tank. It was their intention and had always been under [Wernher] von Braun to eventually not only have responsibility for the booster but to take over the manned functions as well and crew training.

They built this water tank under the guise that they needed it to evaluate their systems. At one time the Skylab part of the program was going to be at JSC, but Headquarters, because of von Braun and other political activities including the senator from Alabama, decided that we should not only have Marshall responsible for the engines and the booster, but they should have a role in the manned activities as well, i.e. being taking an S-IVB tank and putting partitions in it and letting them do some human engineering, human factors in it and build some of the crew interfaces, i.e. all of them if they could get away with it.

So where did that leave JSC with crew training? Well, because there was no Skylab simulator per se, although you could come up with a panel that was in the multiple docking adapter that controlled the environmental functions and the electrical functions in the vehicle, and you could build another panel that simulated the control of the optical telescope. Those were what I can say were very minor simulations. Of course we still had to have the AMS [Apollo Mission Simulator] because we had to go up and rendezvous and dock with the multiple docking adaptor that was the interface with the command module for the Skylab.

It was very interesting. The crews, apparently it didn't make any difference to them where they trained. They didn't care whether they were at Marshall, in a 1G trainer, or in the neutral buoyancy facility, or where, just so they were getting the training. Well, I and Deke felt that JSC should be the central repository for crew training, so I prevailed on Reg Machell and Kenny to have a prototype flight vehicle, in terms of the S-IVB stage that Marshall had more than one of, made them outfit it with all the functions that they were going to put in the flight vehicle which they had intended to use for their own purposes to train the crew, and we put that on a barge and shipped it here. That's why you have that vacated barge dock down there at the end of the road.

So we had the Corps of Engineers come in and dredge a channel all the way from Kemah up there and built that boat dock, and then special GFE and loaded it off the barge, rolled it down the road, and tore down the side of Building 4 there and I put it in.

We got rid of the—I can't even remember what the trainer was in there that I tore out to put that in there. I wonder what was in there, probably a bunch of mockups, mockup trainers that we'd had in there before for the Apollo Program. But we took that in there and then had to disassemble that S-IVB in sections to get it in there and to stack it up and then built the walkway. That's the same system that's over in Space [Center] Houston right now, or whatever that thing's called.

Butler: Yes, Space Center Houston.

Grimm: Yes. So I had that built. Then I had [The] Martin [Company] get approval through Kenny again, and Machell, to build the multiple-docking adapter. Then I had a mockup built—and I forget who built that—of the Apollo Telescope Mount [ATM]. We laid all that on—the multiple-docking adapter and the Apollo telescope, laid it on its side, and then the Skylab vertical. There was another piece in there. What's the other piece? Maybe that's all there is.

But, anyhow, we put all that together in that high-bay area there in the south side of Building 7 and used that to train. We were running into the—and we still had the water tank problem at Marshall. The crew went down there a lot to train. All we had was a little twenty-by-thirty-foot water tank in the back part of 7 that I had moved up from Building 227, which was in that building prior to the Gemini docking training. We disassembled it and moved it up to Building 7 and we were doing in it during the Apollo Program, during all of our EVA neutral buoyancy, EVA training and so forth and Gemini training.

But obviously that thing was not suited for Skylab activities and crawling out on the LM telescope mount to do the film replacement and all of those sorts of things that we had to do on Skylab. So I started the process of getting another water tank, which took me several years, which we put in the rotunda of what used to be the centrifuge. That worked for the later part of Skylab and ASTP [Apollo Soyuz Test Project] and some initial stages of the Space Station development work. Now I understand they have a huge one out at the Sonny Carter [Neutral Buoyancy Laboratory] thing, which that building was built at one time to be the staging area for the ISS. When that went down the tube, they needed to figure out something else for it. So that's when they put the full-sized facility in the ground there to do the neutral buoyancy training for the ISS. Of course Marshall is out of the loop, I think, now in terms of their water tank.

So that's kind of a little history on mockups and trainer.

Butler: There certainly have been quite a few different mockups and trainers and simulators all along the way.

Grimm: Oh, yes. We had a bungee trainer at one time in Apollo 7 so that the guys could bound along like they did, and they were by springs. You heard about that, the 1G Bungee trainer that was in Building 7?

Butler: Heard a little bit about it.

Grimm: Of course we had the air bearing system that Ed White tried to train on for his [EVA] device, which is now that same air bearing thing which a guy by the name of Johnson came up with. But those [steel] blocks that I put in storage at one time and then moved those over to that building, whatever that building is called now, where the big mockups are at.

Butler: Building 9.

Grimm: They used that air bearing facility there. It's the same facility that we had built back in '64 so that the people could stand on pads and move themselves around with this little nitrogen thruster that Ed White used to do his EVA.

So almost everything we did at NASA was part-task. Part-task here, you'd break it up into little-bitty tasks and try to build something that would do that. Very seldom were we ever able to put something together to simulate the whole thing from end to end, that included our fixed based simulators, the Apollo simulator, AMS, the LMS [Lunar Module Simulator], Skylab simulators, Gemini simulators, they were all part-task in one way or another, just like the LLTV.

We had a simulator for the LLTV in Building 4, a little fixed base, so that the crews could get used to all the controls and displays. It had an old B-52 visual system put together that I got from Hill Air Force Base [Utah]. We modified that and scrounged some analog systems to drive the display so that when the crew made an input the picture changed and they could flip switches and go through all of the motions they needed to get familiar with it. They spent about ten hours in that simulator before they ever got out to fly the LLTV. The LLTV [simulator] was a simulator's simulator's simulator. [Laughter] So that's what we had there.

Butler: But it's what you needed.

Grimm: That's what we needed.

As a part of the Skylab, we finally got the trainers built and in and then we started working with all the experiments. So all of the experiments, instead of Marshall doing them down there, we did them all at JSC in that Skylab trainer that I had moved over. The same thing was true for the multiple docking and after then for the LM Apollo telescope mount.

So we had a number of experiments which my organization was responsible for. One was the flying jet shoes. Another one was the backpack, where the crew flew inside of the Skylab. I don't know whether you heard of that or not. And a number of others that I and people I had assigned in my organization were responsible for managing those activities. In addition, of course, we had a lot of the medical folks had a number of experiments such as the treadmill, the Lower Body Negative Pressure [LBNP] device and, of course we had all the food servicing things.

The crews actually went through living there, except for sleeping in the hang-up bags. So our mockup was very representative in terms of practically everything that you could do in 1G. Remember, we had a window that looked toward Earth where you could attach a camera and we did those experiments with wooden ones and with a real one, but without the weight in it. We had an airlock that you'd expose to the sun side, where you put out that. We had all those experiments. The crews actually went through the process of unstowing those out of the containers, putting them together, going through the protocol as if they were actually doing it.

I actually had a team of people go through every experiment and every living function in the Skylab, in the MDA and on the Apollo telescope mount, had it filmed professionally and gave Kenny a set of the film. The program at that time had a set of film, so he had two sets of film. There is another set someplace. [Laughter]

So those are some of the things we did on Skylab. We worked with the crews when they had to go out, mainly to Martin here to get on their simulator, which we strapped a guy on and then he could fly his hand controllers and simulate the same thing as his backpack, which was a precursor to the EMU [Extravehicular Mobility Unit] backpack.

Then about the same time as doing that, we started to work on the ASTP mission. The people in that workshop that I had told you about that we had built a mockup of the ASTP so that the crew could practice hooking the vehicles together and going through the airlocks, opening the hatches and going into the Soyuz and so forth. So that was the main function that we had with that one, plus the experiments that the crews performed during ASTP. There weren't a lot of them, but one of those experiments was managed by the division that I was in later when I moved over to E&D [Engineering and Development Directorate], which we had responsibility for, which was to find out if there was a third oxygen molecule in space, I think was and I'm not sure about that. That doesn't sound, but it had something to do with the third spectral line of an oxygen molecule in space which is validated by this experiment. So there were a number of experiments developed for ASTP, and four or five of those experiments were managed by my organization that I was involved in over experiments. This was in addition to E&D.

Butler: In all those planning for the experiments, both on Skylab and on ASTP and the training for them, were there any surprises as they were training or trying to, as you said, unstow it, put it together, put it up and running, that prompted any major change? Or did it—

Grimm: No. You work out, as I mentioned earlier, just like I and other people in my initial organization went and flew simulations and then we developed the procedures and we changed things and manipulated them and changed hardware and so forth until we were reasonably satisfied we had something that was workable. Then we'd bring the crews in and the crews would add those inputs.

It was an evolutionary type thing so that I don't think we ever ended up with something that was a total flop or that we had to completely redesign. It was more of incremental changes that we made right up to flight on a lot of things as we learned more and more about the experiment and how the crew could operationally interface with it and control it, whatever it was. That's how you do a reasonable program. Not something typically like we did with LLRV/TV where everything was rush, rush, rush, where we never had enough resources, never had enough people, never had enough money, never had enough time. But we did it, and without loss of life, which I think is a big credit to the program, considering how risky it was.

So those were the major things that we did in Skylab. About that same time, the organization that I was in, flight crew operations director[ate], flight crew support division I was in, was getting very large. Some people…thought it was getting too large for Warren [North] to manage. We had 650 people, which is a large organization. We had two assistant division chiefs, Pete [Carroll H.] Woodling, Jim [James W.] Bilodeau, and we had myself.

So it was decided at the director level that we should split the organization into pieces that were presumably more manageable, although all they did was take Bilodeau's organization that he had under him and his branches and that became a division. They took Pete's organization that he had with all the simulators and made that a division. Then they took all the stuff that I had and made that a division, and moved Warren on staff as an assistant director to Deke. A lot of politics [were] involved in that whole machination that went on, and I'm not going to address that.

But it became a procedures division which Bilodeau had, a training division with Woodling had, and the flight crew integration [division] which I had at that time. That was probably at the end of the Skylab and at the start of the ASTP as I recall. That went on for a couple of years.

Skylab program was interesting in one respect in that on the first launch we had the micrometeoric shield ripped off and in the process it ripped off one of the solar rays and pinned the other one down without ripping it off, pinned it down with a piece of shrapnel of the remainder of a piece of metal had hooked the wing so that it couldn't fly out and then the solar rays deploy. With the use of some scanning cameras and tracking cameras at the Cape and some classified Air Force cameras on the ground and otherwise, we were able to see exactly what was holding that wing down and the nature of it to the extent you could see it in actual size, which was a six-inch piece of metal holding that wing down.

So one of the people in my organization at that time, because now all of the GFE equipment had been transferred under me as part of this reorganization, which meant all of the cameras, the pencils and paper and film, you name it, anything that had to do with crew equipment excluding the suits was under me in this division that I had.

So Jim [James A.] Taylor, who was a section head of one of the crew equipment sections, went down to Sears and bought a pair of tin snips and modified those to be able to work them with a gloved hand and had Pete go out and do a little experiment to see if he could work those things to cut that material, because we got samples of the same material that held that wing down. That was the instrument that he took out, or the tool he took out, when he went EVA to—that was the first thing he did when we finally got up there was to go out and cut that piece of metal loose and then stand back because that wing went out and then the solar ray deployed.

Of course the next step was because we lacked the—the micrometeoric shield was also our sunshield. Now we had no sunshield, so the telemetry said that the temperature in the Skylab was getting to the point where it was going to ruin all the film in the film vault that we had up there. All the film was going to be sent up at one time in this big aluminum film vault. It was supposed to last for the thirty-day mission, the sixty-day mission, and the ninety-day mission. It was getting to the point where the film would be hazed and be unusable for any scientific work if we didn't cool that compartment down. So one of the things we had to do was to figure out a way to put out a sunshield that would be over that vehicle in some way.

One of the men [Jack A. Kinzler] over in—I guess he was the division chief of tech services, was a handy guy, I think in a lot of respects in terms of mechanical things. He came up with a device, which was basically a collapsed umbrella, only a big one and sort of like a fishing pole so that once you pushed the umbrella out and extended it and the fishing pole, you could just continue to go out.

However, the problem was that what were we going to hook the fishing pole to and how were we going to get it out there? The thought was that we were going to have the crew go out and carry this as a bundle under his arm in EVA and figure out how to string it from some point to some point to some point to give us the shade that they needed.

Because the sun-looking airlock was sun pointing, and we had our trainer unit there but Marshall had a prototype unit that they used to qualify that system with for a number of experiments that got put out of that airlock area. This box was about yea long, six feet long, maybe a little longer, and so I told Kleinknecht that Marshall had a box that was the prototype that with a little finagling could be flight qualified since it'd gone through all of the environmental testing. If Kinzler could shove all of his umbrella device inside of that thing, we had a pole that was already made with a seal where this pole could be hooked together and shove the experiments out the door.

Well, instead of that, we would pack Kinzler's umbrella expandable device inside of that package, and if that thing would fit inside the command module, we could launch that. They could take it up, hook it up to the airlock, and without doing an EVA, actually extend this umbrella out and it would automatically expand like this automatic umbrella and we would all have our shade. They took me up on my suggestion, and Kinzler worked his part of it and I worked the other part, the box, and the stowage with my people. And that's what we had on Skylab as our first shade.

Now, that thing was pretty flexible and the sun was getting to it, because this was done within a thirty-day period. He was supposed to launch shortly thereafter, you know, Pete was, and his crew after the Skylab was put up in the next three days or something like that, maybe a week.

But in any case it was a short period of time. It turned out to be like thirty days and we were approaching the peak limits on a bunch of things in the Skylab. That device served for that thirty-day mission, plus I believe the thirty-day gap that we had between the first and second missions. The second group went up and strung another one, manually, EVA on the outside of the other one, and that then worked for that mission plus the thirty-day gap plus the next ninety-day mission on there.

So there were a number of little things like that that we did on Skylab, but that's the one that comes to mind as an interesting little project that I was involved with along with a number of my people.

Butler: Certainly a very important project to keep Skylab functioning.

Grimm: So you said here, Skylab, what other responsibilities? I think I've covered the number of things that I've done along the way.

Butler: Sure, you covered Skylab pretty good.

Grimm: Then you've got here, after Apollo XIV I became the chief of flight crew integration division. I think I've told you about it already the work in there and what the primary responsibilities were of these numbers of teams that I had and these people who did these various different things.

Butler: You covered that pretty well, I think. We talked about the Apollo Soyuz.

Grimm: You said how was it similar and how was it different, and I think we've covered all of those things.

We worked, as I've talked about, on Skylab's experiment integration package. I think we've already covered that and how I got involved in it. The main reason I got involved with it is because I will take credit for the fact that we did get the training transferred from Marshall to JSC, that 1G training. We already had the simulator training, but had I not pushed that all of that work would have been at Marshall and the roles and functions of the centers at this point in time might be different than they are now. That's just conjecture.

Butler: I think it would have changed a lot.

Grimm: They certainly would have gotten a lot of experience. Of course, they did anyhow because Marshall put together the floors with that grid pattern. They put together the handrails. They put together the food lockers. Of course, JSC actually did the food. But they put together the sleeping activities, the trash dump, the water tanks, the storage lockers in the upper area, those sorts of things. They did the environmental system. They did the scrubbing system for the CO_2.

Butler: The lithium hydroxide cleaner?

Grimm: No, the molecular sieves. The molecular sieves is what we had on Skylab because that was a continuing thing. You had two banks of the molecular sieves, and you would work one bank until all of the sieves were filled with the little CO_2 molecules, if you can imagine them as little jellybeans that wouldn't go through the hole while the oxygen did. So that scrubbed out the CO_2 until they got full, then you'd close that off to the interior and you'd open up the other one and operate in. Then you'd open this system that was presumably saturated and vent it to the hard vacuum, and it would suck all the CO_2 overboard. Then it would be cleaned, and then you could have that one ready as your alternate scrubbing system for CO_2. That's a simplistic way of describing that, but I'd even forgotten about that.

Butler: Interesting system.

Grimm: So as I said, they were involved and got a lot of initial work done in human factors, assuming that they were going to continue in that. But there were some changes in management, changes in Headquarters and I think that for the most part has sort of gone away, that JSC is still recognized now as the crew interface and the human interface to the spacecraft, and Marshall is mainly the engines and the module and so forth.

But there was a lot of politics involved, and they had a very vocal spokesman for themselves by the name of [J. R.] Thompson down there. He was a branch chief at the same time that I was probably equivalent to him. He was a very vocal guy. Later he became the manager for the Shuttle engines for Marshall, and later when the Challenger accident occurred he was on the accident investigation board. He had by that time retired and he was working for Orbital Science Organization, he was the vice-president, I believe. Then he went and either chaired the board or he was next on the board. I'm not sure which one he was.

Then he came back to NASA as the deputy administrator on that job, and now I forget where he's at. He's not in NASA anymore, but he was a very aggressive individual. He was even more aggressive than I was. I always had in a backhanded way hand it to him. But he had a lot of support for his position from his management down there, and I sometimes had to really work on that. Reg Machell, in the program office, and I don't know whether you've interviewed him or not.

Butler: Not yet, but we're hoping to.

Grimm: He was not what I would consider an aggressive individual and had to do a lot of convincing. One of the things that we started on was this attachment to the end of the manipulator arm that the crews stand on and has a pedestal with all the tools and so forth. Well, he thought that was a bunch of crap. We had to work on him and the program manager for damned near a year until George Franklin finally showed that as just a foot restraint, and then we kept adding on to it and adding on to it.

Now it's a mandatory piece of the system, so that the crew doesn't always have to tether himself and control his body and all the other problems he has when he's handling large pieces of equipment out there, such as the Hubble Telescope replacement-type things where it gives the crew a very stable platform and all of these tools at hand to do that.

So those are a number of things of the types of things that we when we were in the crew area, and even when we weren't in the crew area, when I was in the engineering area and George was still in crew organizations, we really had to fight for all of those things. At the time people would say, "That's not very important," or "We don't need that." Like, "We don't need a window in Skylab that's optically clear." But once it was and we put a large camera in that window and we took all those neat pictures, we had the same problems coming up in the Shuttle.

"You do not need an optical window in that side hatch." "Well, it's been very valuable." Guess what, comes around to the ISS, "Who in the hell needs an optical window in the ISS?" Well, there is one now, but it took a lot of work by some very dedicated people to get an optical window in there and it's going to be very, very useful for a lot of different reasons.

That's probably one of the criticisms that scientists have of NASA as an organization, because they think it's engineering oriented and engineering run and we don't listen to the scientists. In many cases that's true, and in many cases it's tied to cost. Other times, it's tied to whatever whim the particular program manager has at the moment. A lot of people would be surprised to see that many of the astronauts, even just the operational piloting types, see a value in a lot of those things that the program doesn't see, mainly because it's schedule or cost that's involved.

Just like the inward-opening hatch. We, I say we, I wasn't involved. It was Warren North and the crew looked at that activity and went and pitched to the program manager, whose name I have never said yet and I'm not going to, said, "It's too costly and it's too much of an impact to put an outward-opening hatch in there." So we didn't, and what it cost us was three astronaut lives and a hundred to two hundred million dollars. It's just in rework, not considering the rework and the module we lost and everything else.

So people attach price tags to things, perhaps when they don't have a good appreciation for crew interfaces and the safety of operation. That was the thing of course that I and my group of people always looked at. We always looked at what makes it operationally easy for the crew to work with this device, what do we need to make it safe for the crew to operate with this device, and thirdly, what do we need to do to this device to make it functionally useable or scientifically useable and get data back. It was kind of in that order, and that's the way you have to put your priorities. But sometimes we had management who would just worry about cost and schedule and not worry about these other three items that I've talked about.

Where are we here?

Butler: Talking about and mentioning the window in Skylab and such, was this when you began to become involved with the Earth resources work was during Skylab and as you were looking at some of these experiments? Or did that come later?

Grimm: Your question is how did I get involved with experiments, or how did I—

Butler: Specifically with the Earth resources package?

Grimm: I was involved with them because, since my function was the flight crew integration function. Then I had a number of people assigned to follow every experiment that the crew was involved in. In most cases before the crews ever got involved with those experiments that our people had looked at those experiments, evaluated them, gone through preliminary procedures, done the experiments themselves, whatever that was, and developed the first time lines and procedures before the crews got involved, just like we did in the original one I was talking about when we'd fly the simulations and evaluate them and develop procedures and so forth and work out some of the bugs and then get the crews involved for their evaluation.

We did the same thing with all the Skylab experiments. So as I said, I had people assigned to groups of experiments, medical experiments, solar experiments, Earth looking experiments, interior experiments and so forth. So that it was their responsibility to make sure that all the safety aspects were looked at, all the crew aspects were looked at, all the time lines were developed, the procedures were developed so that we could give all those things to Tommy Holloway's group at that time to integrate into a set of procedures in a total time line for the missions to accomplish.

About that time…the FOD got disbanded, the FCOD [Flight Crew Operations Division]. I don't know whether you heard of that or not.

Butler: Some.

Grimm: The way you say that, I'd be really interested to know what you know about that, but let's let it suffice to say that it was disbanded. A lot of political overtones associated with that. As a matter of fact, it was specifically political overtones that caused that disbandment of the flight crew ops directorate.

As a result, [Eugene F.] Kranz, I believe, was the director of FOD at that time, because I believe [Christopher C.] Kraft [Jr.] was the director. So they wanted all the mission planning and procedures work in their organization, mainly because they saw how good Tommy and that group had done. Because it got to the point where FOD used to be at the consoles and trying to do the flight planning, but when they got in a pinch they'd ask Tommy to come over and sit in and to assist them in their work during the mission, and it got to the point where he actually had his own console there and operated pretty much as a right arm of the capcom, in terms of passing out procedures and so forth although Tommy was very good at working with the FOD systems people. So that's how that organization for it went.

The training organization, FOD didn't want the training organization with all the simulators and things like that. As a matter of fact, they were almost trying to build their own simulator organization themselves and they accomplished that to some degree. Again politics were involved. But for whatever reason, and I'm not even privy to this one, that they didn't want that organization. So it was going to stay in a flight crew organization.

My organization was split in three different pieces. One part stayed with flight crew. One part went to another division in E&D. The other part went to the organization I eventually went to. I have to be careful how I say this. There was some discussion about who wanted what, who wanted who and why.

I will say this, this is the only part I will say, is that in my discussion with Max [Maxime A.] Faget, he called me over and said, "I've got a couple of openings in my organization as division chief, and I'd like for you to consider it." I asked him why was he considering me because the whole thing was separated from my organization. His comment was, he said, "Well, I have watched you over a period of time, and I have had other people evaluate you and I've polled all my division chiefs about what they think about you and others." His comment was that their total evaluation was that I was probably the best manager at JSC. That was more than he could say for some people, including some in his own organization, and that he would be glad to offer me one of these several positions if I wanted to take them.

Well, it is kind of like being hung out to dry when your organization's disbanded and parceled out and you don't have the benefit of any input. So I went to talk to this one other gentleman in another organization under Max and I decided that there were better things in life I'd rather do than fight with that individual for the rest of my career, and so I took what Max considered to be the lesser desirable option. I took this other division because he was moving the division chief out of that organization for reasons I won't go into, and said, "That organization is yours." But he said, "It's a hell of a mess, and you do whatever you want to with it, but don't bother me with it."

That's just the kind of direction I like. So in about a year's time I reorganized that operation. It was split in five different buildings at JSC including a piece that I was eventually to get and bring over that I had in my old organization, which was all of the GFE crew equipment excluding suits and camera gear and all that sort of stuff and the design group which I really wanted because you can do a lot of things if you have your own design group. And some of the shop equipment. I wasn't able to keep it all. That went to another organization.

But over this next year's period I physically booted two other divisions out of Building—I forget which one—17, 16, it was two buildings over from the auditorium.

Butler: Okay, it's one of those.

Grimm: Maybe 14, I can't remember. But some of these people wouldn't move, so one weekend I told them if you don't get your people out of this building by such and such a date, the next day when you come to work, all your stuff will be out on the lawn. They did not believe me and that's exactly what happened. So I vacated their desks, their books, their everything, and stuck it outside. Of course, I told Max what I was doing and the assistant [directorate] chief that he had, Bob [Robert A.] Gardiner.

Bob Gardiner said, "I don't care. Go do it." He said, "These people need to get their facilities aggregated and I've been telling them for months…what to do, and they wouldn't do it." So of course they went bitching to Bob, and Bob said, "Tough, you're out of there. So you figure out what to do with your stuff now."

So here I had five groups who were pieces of five previous organizations that had never been integrated. So I spent a year putting them together, integrating them, getting them to talk to one another, reorganizing them, having them all in one building. At the time I ended up with a building that had the accumulation of stuff from Gemini, Apollo, Apollo-Soyuz, had the measurements systems and calibration lab for the Center where all the instruments for the Center come in. I inherited that when I inherited this lab. I inherited an IR [infrared] group. I inherited a microwave group. I inherited the GFE group, and I inherited a couple more groups and all of the aircraft Earth resources operations.

None of it was working very well, and the building was full of obsolete equipment. So I spent a month inventorying. I don't know if this is of interest or not, and you can eliminate it if you want to, but I'm just going to tell you anyhow. [Laughter]

Butler: It's interesting, absolutely.

Grimm: So I just spent a month inventorying the entire building and making everybody in that building justify every piece of equipment that was in it. When they couldn't justify it, I said it's surplus, and I called up facilities people. I can't remember who handled all the surplus equipment on the site, back in the back forty someplace. L.C. somebody, was the division chief. I surplussed all that equipment. We physically ripped it out of the building and set it outside and had somebody haul it away. The rest of the equipment that we needed, which were the small vacuum altitude chambers and all of the MSCL equipment, measurements and standards equipment for calibrating all the equipment on the Center, I aggregated it back in the high-bay area. Then we put people in all those rooms.

Then the next thing I did was it was such a dirty, filthy place that I asked facilities to paint it. They wouldn't paint it so I had my support contractor buy fifty gallons of paint, and over one weekend we painted the whole building, which irritated a bunch of people. Painted the whole building, and then I appropriated carpet and without authorization from facilities carpeted and made two carpeted rooms and put pull curtains between them so that we could have different meetings at different times. I put in a music system so that we had elevator music in the offices, segregated smokers, and in general cleaned up the place, raised morale, hung nice pictures on the walls of the space program, and so forth.

That was just the start of getting things going, and then I started working on the technical aspect of the program. Most of the programs were in fairly good shape except for our microwave program and our infrared Earth resources operations. At that time I still had responsibility for the Skylab experiments. That division had some of the responsibilities, and then I transferred the rest of them over. So we continued to monitor or build and develop experiments that were to fly on Skylab and subsequently ASTP with a group of people.

In addition there were five airplanes at Ellington. We had a C-130. We had a P-3. We had a Lockheed Electra. We had a U-2 and we had a B-57 and we had a couple of helicopters and then we had some ground systems. All of these were to support an Earth resources division that was over in the science directorate and Goddard [Space Flight Center, Greenbelt, Maryland] and Purdue University [West Lafayette, Indiana] and a number of other universities in developing the science of predicting crop productions and many other things, salt infiltration, bugs, crops growths, area growths, forest, infestation of bugs, inventorying commodities. All of these things were done with ground-based systems on the ground to get ground truth, helicopters for a little bit higher, C-130s and P-3s for a little bit higher, up to 30, 40,000 feet, and then the U-2s and the B-57s for up much higher.

Then the ultimate thing was to support Skylab when it flew and to take all this data and aggregate it and process it. I actually then built a computer processing facility that hadn't existed before I arrived, ended up with one of the biggest ones in the Center, except for the main processing mainframes over in ISD [Information Services Division], so we could process most of all the data that we accumulated on all of these vehicles nonflight, and then also some of the flight data we processed from Skylab. So it was a top-to-bottom iteration of data and then correlating all that data top to bottom so that then you could do an evaluation of whatever resource it was you were looking at.

The problem with all of that was that a lot of the equipment we had was obsolete, wasn't well organized, we couldn't make changes fast enough to support different investigators and science investigators from all over the United States and universities and so forth, or even our own. Everybody was pretty parochial so I made the decision that we were going to modify the C-130 and the B-57 and the U-2 and put in a system where we could do like a snap-together system.

In other words, if I wanted to fly a completely different mission, if I wanted to fly a camera mission today and I wanted to fly a UV [ultraviolet] or IR mission tomorrow on a B-57, I wanted to do it. Well, it took us maybe two weeks to do it before or maybe a month, because these things were hard mounted in the payload bay of the B-57, as an example, and in the nose.

So what I did was I had an interchangeable nose built for the B-57 and put different instruments in each nose, so that we'd run the dolly under the nose that night, take the dolly off, stick a new nose on, had a patch panel built in, run new cables back to the payload bay where the computer was and where our data recorders were and repatch it and we're ready to go as for the nose.

All the instruments that fit in the payload bay underneath were hard mounted again. I said, "That's never going to hack it." So I took the doors off, I had tech services build me a bunch of pallets, one six-foot pallet and some three-foot pallets and hooked up a cable system so that with a dolly we could run under there, hook two cables or four cables to the dolly, crank it up with a crank, cinch it in place, do a patch panel, check and patch all the instruments into the patch panel and to the recorders, and we're ready to fly the next day.

So we did that on the B-57, did the same thing on the U-2. We ripped out all the fixed installations on the C-130, built a new nose on the 130, put in a scanning radar system, which they said couldn't be done, and ripped out all the fixed installations inside the cabin, put in rollout cabinets with all of the control systems and recording systems and crew stations inside the C-130. Put all of our instruments on quick hangars that we could stick under the tail or on the ramp door or on the wings so that we could reconfigure those airplanes totally, and we could be flying in South America yesterday, land at JSC that night, and one day later we could be in Alaska on a completely different mission.

So that was some of the fun and games that I had with that organization in the experiments systems division that I had formed and so forth. Another thing I didn't like is when I first started out working for NASA, we had NASA engineers doing all the front-end work. So we knew what was going on all the time and used contractors to do the handwork, the hands that you didn't have because you didn't have all the hands for resources.

Over a period of time, we got away from that philosophy across NASA, where we hired contractors and we became managers. I don't include myself in that category. Our people got upgraded, and instead of flunkies, 9s, 10s, 11s, 12s, 13 grades, they become 14s and 15s. They got to the point where they didn't know how to do anything except monitor contractor's work and monitor the contracts. So we were gradually losing our technical capability. That's not just in my organization; it was across NASA. It is across NASA now even more so than it was when I was there.

You're losing a lot of old heads, or you've already lost them, like me over a period of years, and they haven't been replaced because that philosophy is not there. I say.

Butler: You're not the only one to say that.

Grimm: So what I told my guys in my organization is you're going to use your contractors to implement stuff, but you're going to do the initial brainwork, you're going to do the planning, you're going to do the budgeting and you're going to see that it get built and we're going to use the contractors to do whatever we want.

So I gave several of my branch chiefs this directive. I said, "I want you and your people to use this design group that I have here, and I want you to design your side-looking radar. I want you to design your IR instruments. I want you to design your camera systems. I want you to design all of these things that we're going to do and put on these airplanes to fly, because I want you guys to have that knowledge back in this organization. If you don't agree with it, then you're free to transfer to any organization that you like in the Center or out."

Some of them took me up on it and they left. That was fine with me, because I didn't want those people in my organization. I hate it when a contractor can do or did work that I don't understand, and not only that, that I'm not in a position to understand because I haven't been following it from a technical standpoint. I didn't think that's what NASA was organized to do. We had some other gentlemen at the top like [Daniel S.] Goldin who think differently, and I certainly don't agree with him on many of his philosophies but he's the administrator and I'm retired so it doesn't make a hell of a lot of difference what I think as far as he's concerned.

Speaking of him, he's gotten rid of a lot of people who have my same philosophy and put in people who at least espouse his philosophy or follow his line of thinking.

Butler: It's certainly changed a lot from the early days.

Grimm: Oh, yes, it has. I think you can tell that from the way I've described what we used to do when we first started in the programs way back on Mercury. That was one of the first tasks I had when I came on the program, other than the LM instruments, was to work with John [H.] Glenn [Jr.]. It was doing a manual reentry simulation of Mercury, and that's an interesting thing to do.

Scared the hell out of me when I did that because I said I don't want to be an astronaut if this is what you've got to do. But it was one of these things where you actually got in there and did the engineering yourself and understood what was going on. Then when the crews flew, you understood precisely what they were doing, which I think is a good thing.

So I think I've talked about this, didn't I? I was recognized for creating a rack system to carry the experiments on the C-130.

Butler: Yes, we talked about that.

Grimm: But that applied for the same work on the U-2 and the B-57. Interestingly enough, after we did all these things and were a tremendous success, including using that side-looking radar in the nose of the B-57, we mapped all of Alaska. They are making a big deal out of this one that just flew where we mapped the world, but I mapped Alaska way back when, when we didn't have GPS [Global Positioning System] or anything else with side-looking radar. I actually did the data analysis and data correlation and distribution of the data in my organization. That's why one of the things that I did to develop this large computer lab that I put together that people said I shouldn't have, but they said that about a lot of things that I did.

But anyhow, after we were a tremendous success doing this Earth resources stuff with the airplanes, there was a lot of complaints by Langley and especially by Ames [Research Center, Moffett Field, California] that we were doing work that they should be doing. Then Langley got into the act with Wallops so they sent the U-2 and the B-57 to Ames. They sent the C-130 to—where did they send the C-130 to? It might have gone to Ames, too. They sent the P-3 to Wallops. I forget where they sent the helicopters. Anyway, Kraft agreed and they disbanded our total organization there, and on to bigger and better things.

At the same time, we were working on Shuttle. Since we weren't working Earth resources and Skylab and ASTP had come and gone, and I started looking at what Shuttle payloads we were going to have. We were assigned by the program office to develop a small organization that had the proper clearances to support some national organizations and work with Headquarters and those organizations to develop and fly those payloads on Shuttle.

In the process of looking at this activity, I thought, "Well, now, why can't we fly a package that's related to things of interest to OSTA, as an example, or OAST?" I'm not sure those are the same; they're not, are they? OAST is Office of Aeronautics and Space Technology. OSTA is Office of Shuttle Transportation; is that right?

Butler: It sounds right. I don't, unfortunately, have the acronyms here.

Grimm: I can't remember. It's been too long.

Butler: I can check that. We have copies of old phone books, so I can pull those.

Grimm: Okay. So anyhow this had to do with Earth-looking experiments. Not necessarily Earth resources, but Earth-looking. I got our guys down to brainstorm. We sat down in a meeting and we brainstormed a number of things over a period of weeks. I said, "Well, why can't we get a pallet," although Marshall had dibs on the pallets again. Here's Marshall again getting into the act to placate—and of course at that time they had somebody at Marshall who was at Headquarters in a position to divert work to Marshall. So they had the pallet design responsibility.

But I said, "Why can't we get a pallet, and why can't we develop a package of experiments that looks Earth-looking, such as the large-format camera as an example?" Good story about that. But that large-format camera, I don't know whether you've heard of it or not, that did fly, twice, but there was so much politics associated with it because it was part of the K[H]-11 Blackbird satellite program that was classified. Lockheed built the system and launched it for a national organization.

We sort of slid this camera out from under the umbrella of classification and said we wanted to fly it. Well, I had two problems. One was the classification. The second one was that the gentleman in Headquarters who we had to get authorization from by the name of Pitt Thome had committed to JPL [Jet Propulsion Laboratory, Pasadena, California] that we wanted a thematic mapper [TM]. This camera would conflict with this thematic mapper that he wanted to fly. So his attitude was we were going to kill this camera so that it wouldn't deplete the funds or interest from the scientific community. I was just as determined that we were, and so was the project engineer I had on the project named Bernie Moberg, who was a fanatic when it came to this camera.

This was a 24-inch focal length camera, which probably doesn't mean much to you. The lens weighed 600 pounds and almost two feet in diameter and had the capability, was totally color corrected, no streaks whatsoever, and could take phenomenal black and white, color and UV color corrected pictures and very sensitive black and white with a resolution if it were allowed by national organizations of less than three feet at that time, which is even pretty good now.

So that was a big problem, so I fought the battle of that at Headquarters. Finally we got approval to go build that system, which we did and we flew it. Then Pitt Thome deep-sixed the system, said it doesn't have any scientific merit, although we mapped two-thirds of the world between the sixty-degree latitudes or fifty-seven fifty-seven that we were flying in Shuttle. Maybe not quite that high. It was a good portion of the civilized world that we mapped when we didn't have cloud cover, and with different kinds of film.

It was all stereoscopic because this was a moving Shuttle camera where the film moved while the vehicle was moving. So you got a picture here and then so many seconds later we got another picture at a different angle because we had it on a rocker so that the camera would look at the same spot and therefore with two pictures looking at a different spot from a different perspective, you get stereoscopic perspective and enhanced altitude of the landform.

As a result of our pushing, over his objections, we sneaked a few questions to a friend of ours, a legislative aide on one of the senators who was interrogating this gentleman on a Hill review and got him to admit that this was a good thing to do and got the funding for it that way. He accused I, and Bernie, of manipulating the system, and if he ever proved it then we were fired on the spot.

Fortunately there's never been any verbal proof of that until right now, [Laughter] which is still kind of funny.

Butler: Well, it certainly sounds like it was a very useful system.

Grimm: It was. Of course that thing is rusting away down at New Orleans, at Stennis [Space Center, Louisianna]. We spent twelve million dollars on that system, fantastic system, and because he hated it so bad, it never flew again after two flights.

That's kind of the breaks of the game. But in any case, we continued and that was one of the things that flew on its own separate pallet that we finally got it on. Bernie managed the camera and Curt [James C.] LeBlanc and Jack [Jackson D.] Harris managed the pallet and the integration of that pallet with Rockwell and the flow through the Cape.

But the other pallet was actually a pace finder for all pallets that went through the Cape in terms of flow, because our OSTA pallet was the first actual-sized pallet. The Air Force pallet that flew on the first mission was not. It was just an instrumentation pallet with a single post and yeah wide with a lot of instrumentation. Ours was a full-sized pallet.

We actually then worked with other centers and other experimenters to put together the package. We put together the SAR package, the one with the big antenna. JPL found out what we were doing. They went to Headquarters and got Thome to assign them the responsibility for the recorder and the evaluation of the data, but we still had the responsibility to put the hardware together and to get the antenna built by Ball Brothers and to put it in there as a fixed angle.

We had put together the whole package, and I went to Headquarters and sold it not only to Pitt Thome, but to the associate administrator for OSTA and got the funding for it. I sort of got my proverbial you-know-what in a crack because the associate administrator called up [Chris] Kraft and says, "Hey, that was a great presentation that Grimm made up here, and I assume you're 100-percent behind this." And Kraft says, "What are you talking about?" And he told him, and Kraft sort of hesitated and says, "Yeah, we're 100-percent behind it." The guy says, "Good, you got the funding. Grimm's got a good plan there, and you got the organization to do it, so everything's authorized."

About a second after that call, I got a call from Kraft and he said, "Get your ass over here," and he sort of nailed me to the wall because I hadn't gone through the program office, even though there wasn't an interface for me to go through really. So they created an interface. They created an office to manage me while I was doing the work. That made Kraft [happy]—

Butler: That's pretty good.

Grimm: Well, isn't that the way things happen?

Butler: Absolutely.

Grimm: So that manager really didn't give me a hard time. His name was Dick Moke, quite a character to deal with but he was a lot of fun. Fortunately he was a guy I could work with, because otherwise I would have had to kill him. [Laughter]

Butler: Before we go any further, I'd ask if we could go ahead and [change the tape].

[Brief interruption – tape changed]

Butler: We're on.

Grimm: Okay. I'm not going to discuss the experiments in general except to say that we worked with JPL and Wallops, Langley, a number of universities, which were horrific to work with. Not terrific, horrific, because those, quote, «scientists,» the P.I.s [principal investigators] thought that all the money that we sent to them was theirs for their personal use. If they missed schedules or didn't provide it on time, just send them more money and solve all their problems. So we had quite a problem in that. We had to, as a matter of fact, in several cases replace the P.I.'s, which caused a lot of turmoil in the universities because we couldn't get them to provide the work that they had contracted to do.

During the same time that we were developing this first payload for Shuttle, which we did, and as I said earlier it was used as a pathfinder for all the functions at KSC flowing through the payload facility and onto the stack and into the payload bay and so forth. During this same time or about the same time, we had discussed with OAST or OART, I'm not sure which one it was at the time. I think it's OAST, but it might have been OART and then OAST. You know they changed the lettering of those organizations up there. But it was Shuttle experiments, and I can't remember what the Shuttle experiments were called and you don't have it here [referring to pre-interview notes]. Maybe you didn't find it.

But what we did was we put experiments on Shuttle. On one of the Shuttles we replaced the tail pod up on the rear tail with an IR camera. With this IR camera during reentry we could photograph the heat flow across the upper wings and find out what the heating values were at any point on the top side of the Shuttle. We replaced the heat tile at various places on the vehicle with different types of experimental tile. On the wheel-well doors, we actually put a sensor inside the wheel-well to see if we could come up with a device that would give us a measure of altitude because with the plasma flow around the vehicle during reentry, it's very difficult to get an altitude reading and get an update with the radar altimeter on the vehicle.

So there was a whole series of experiments that was funded by and I'll say OAST that we—I was trying to think of the associate administrator we pitched it to. Hinners, I think, Noel [W.] Hinners. He's out here at Wateron [Colorado] at the Lockheed plant as a vice-president of something. But he was the associate administrator. We pitched it to him and we got it approved and then we had to pitch it to the Shuttle program manager and then we had to pitch it to Rockwell and then we had to worry about the integration of all of these experiments so that it wouldn't affect structures, it wouldn't affect loading, it wouldn't affect heating, it wouldn't instrumentation, and all of these things.

Then we finally got our own recording package installed so that we could record the data with our recording package. That, after I had initially pitched it to Headquarters with Max Faget and got it approved…I assigned it to Don [P. Donald] Gerke. Don Gerke managed that with a group of people in my organization, and Don Gerke was my assistant division chief at the time when I had the experiments systems division. He's the gentleman that I said had passed away with a heart attack about five years ago, who went to sleep and didn't wake up, a terrific young guy.

So that was another activity that we did in conjunction with the OAST. Of course during this, about that time after we'd flown that OAST or OSTA package, whatever it's called, and this Shuttle experiments package, some little acronym that it had—

Butler: Do you remember which mission that was on? I can—

Grimm: Oh, it flew on a number of missions. In particular, this tail pod thing was this big bulbous thing on the end of the tail, so it flew on a half a dozen missions of that vehicle, because it wasn't interchangeable with the other vehicles.

Then about that time, Max was getting ready to retire and both his deputy and his assistant director for avionics, all of them were about ready to retire. I think Max was being encouraged by certain people to retire. So they selected two system division chiefs. One was Al [Allen J.] Louviere, who had six divisions that he was responsible for, and I had six divisions that I was responsible for. So they brought us onboard. This was about in '79.

That worked out quite well because Max and Alec [C.] Bond who was his deputy and Bob Gardnier who was his assistant director could go do their things in terms of planning and in terms of thinking of new technical ideas. You know Max is the one that basically invented Mercury, Gemini, Apollo, and Shuttle, the concepts, so he's patented them. I don't know if you've interviewed Max or not.

Butler: We sure have.

Grimm: So you know that, or did he tell you that he had?

Butler: He did. We talked about that.

Grimm: He's a very interesting individual to talk to.

Butler: Yes, he is.

Grimm: But Max always let me do my thing, and I was always very salutatory toward the work that I did. As a matter of fact, in our big retreats he would use me as an example, which really pissed a lot of his other division chiefs off, as the way to manage your organization and get things done, which didn't make me a lot of friends with those guys, but for the most part we got along reasonably well.

So we were selected and then I selected a division chief to replace me and combined two divisions, which was George Franklin who had [previously] moved over [to another division] when my organization [FCID] was abolished. He went over to another organization as a branch chief, and then I took over this [Engineering and Development, E&D] division. So when I went up to the director level, and Al Louviere who was the chief of that division came up as the other director, then George became the division chief there and Don Gerke became the division chief [of my previous division].

Later on, those two divisions were merged again, that had been separated, parts of each one of those had been my organization back in the flight crew days. So they brought them back together again, and then George became the division chief and then Gerke went on staff and then finally to the [space station] program office and into Headquarters later on.

In that position, I was responsible technically for the six as Al was for his six organizations. We were responsible for getting the manning together, the personnel, the budgets, all the administrative actions, the management actions, sitting on change boards, worrying about facilities, getting money from Headquarters, selling programs. He worked mainly on the initial studies. He did probably a couple years' worth of [space station] studies, [and while] he did…all of that, while I was managing sometimes part of his organization, sometimes not, while he worked the initial design and development of the Space Station studies.

Then later that was passed off to Clarke Covington who—have you interviewed Clarke?

Butler: No, not yet.

Grimm: —who was in the program and was in our division and then moved to the program office. Then he picked up the responsibility for the Space Station for a few years until politics got the better of him and somebody else was assigned to pick up that function as the manager of the ISS at that point in time.

So the next three years was mainly a study in transition, turmoil and reorganization. [Chris] Kraft decided that [Gene] Kranz needed more people over in FOD, so we were directed to come up with a list of people that Kranz and his division people could select from. We were supposed to give up out of our 900 and some people, we were supposed to give up like 150 people as I recall. So we rated all of our people. We took all of the people that we had rated, and we cut off the 150 bottom people and gave that list to Kranz and to Kraft.

Well, you can imagine what happened there. The crap hit the fan and Kraft said that's unacceptable. Kranz said, "That's a bunch of blah, blah, blah." So Kraft eventually made us do was give him and Kranz our rating of all these people and then we were supposed to pick a certain percentage out of each bracket, the first five percent, the first ten percent, the first fifteen percent, the first twenty-five, and right on down. Then they could select people out of those brackets that they wanted.

So that happened. That caused a lot of stress and anxiety, if you will, in E&D, both in the people that were getting transferred who didn't want to get transferred and then, of course, in our organization. But by that time Max had left, retired, Alec Bond had retired, Bob Gardiner had retired, and Kraft brought in Bob [Robert O.] Piland. Have you interviewed Bob Piland?

Butler: Yes, we have.

Grimm: Okay. It would be interesting to see what he had to say. But Bob's mandate from Kraft was to whip us into shape, whittle us down, cut our contractor resources, cut our manpower, transfer people and in the end take our functional assistant director responsibilities away from us and put us as titular heads under him as just assistant directors only.

Butler: Interesting.

Grimm: I thought so. I guess the least that I can say about Bob Piland is he has one hell of a big ego. That's probably the only thing I want to say about him.

Then after he had done all this to us, Al and I, I think Al and I started looking at what other options were available to us in NASA because that didn't seem to me to be a very agreeable assignment. I don't believe it was to Al either.

Butler: Understandably.

Grimm: Because that said [our new charter] that we could talk to those division chiefs and we could tell them to do what we wanted them to do, or ask them to do, but in the final analysis they could go over our heads and talk to Piland to decide whether or not he wanted them to do it or not.

Piland had been at Langley and was what I call a fair-haired boy there. He was brought to JSC in another fair-haired position. He went to the Earth resources lab, which is now part of Stennis at New Orleans, ran an organization there, a very small organization there for some period of time and was brought back to JSC to do this reorganization after Kraft convinced Max to retire. So Piland came in with a mandate or a number of mandates from Kraft, which he carried out very well.

So we looked around, and they advertised a couple positions, one at Headquarters, one as assistant director and a director at the center at Dryden [Flight Research Center in California] at Lewis [Research Center in Cleveland, Ohio]. I looked at those jobs and really didn't want to move because at that point in my career I had twenty-seven years in government service plus industry. Then Piland left and got assigned to director of something, because I guess Kraft thought he needed some ass-kicking done over there. I believe that's the way it was and not the reverse.

But in any case, we had a vacancy there. Well, I thought, predicated on our past performance, that either Al or I were qualified to become the director of E&D. In looking at subsequent appointments, I think we were. But we weren't, and they brought in another gentleman by the name of Aaron Cohen from the Shuttle program office. He'd been the Shuttle program manager, which I thought was interesting to go from a program management job to an institutional directorate job. Usually you'd go the other way around if you were doing that.

Aaron came in and Al was exploring his options. I think he left like three months after I did and went to the Space Station program office. But I left, I think, a week or two weeks after Aaron Cohen came in, and for a number of reasons. One, certainly that I should have recognized the politics of selection at that level. There was another reason that I left that I'm not going to go into.

Butler: That's fine. Time to move on to other opportunities for you.

Grimm: Pardon?

Butler: It was a good time to move on to other opportunities for you.

Grimm: That's exactly what I said on my form, I think, that I had experienced a lot of things at NASA and that I was going to explore what new vistas lie out there ahead of me. So I legally, after the fact, and not saying how much before the fact illegally, I secured another job with private industry with a company in Austin, Texas, as a manager in a C3I [Command, Control, Communications, and Intelligence] program.

Do you know what C3I means?

Butler: Yes.

Grimm: I worked in that program for a year, which had a certain level of classification. After a little over a year, I moved into a higher classification program, black program. Do you know what that means?

Butler: Yes.

Grimm: I spent a year at Sunnyvale, California. After that year I came back and on that program I was the program director for all engineering on that program. I came back to Austin and became a program manager on another black compartmented program and spent a number of years there. At that point in time, there was some move afoot, and sometimes I had spent TDY like six months at Burbank on some other black programs. Then they wanted to send me as program manager to Nassau, New Hampshire with another company for assignment for a year, and I didn't want to move back there. So I took a quick retirement from that company and started consulting.

For the next six years I consulted in the aerospace industry, working for Boeing, for Grumman, for NASA, for Lockheed, for McDonnell Douglas, in which I signed contracts of a defined period of time to deliver a defined set of products. When that was through, I was free to go do more work for them if they wanted more work or go take a siesta or go on to another company's work.

To be very truthful, a lot of these jobs I took so that I could do genealogy on the weekends.

Butler: Well, that's a good reason.

Grimm: So my wife and I did a lot of traveling up and down the west coast and the east coast and the south and Europe. So that was a very interesting time, certainly profitable, because when I retired from NASA the pay was minuscule. It's not now. I think the congressmen just gave themselves a raise and the NASA people a raise. I think my grade when I was in was senior executive service, and I think they are getting 125 or 150,000 for that which is doable, decent.

I think we are about through here.

So I continued to work as a consultant for six years and then I had a medical problem. It took me out of the loop for about a year as a result of that. About that time, [George H. W.] Bush cut all consulting work from all the contractors including NASA, he was cutting back on the budget. Any work that had to be done was going to be brought in house, which everybody did to the extent that they could. So the consulting activity just sort of dried up for a few years.

I've been asked from time to time since that point, since '92, to consult again, but they were offering me a consulting fee that's not commensurate with what I consider my experience to be and it's more in the way of what they wanted to pay us like journeyman consulting fees rather than somebody with some expertise building programs.

So during the consulting I worked on the Space Station for Boeing and Grumman up in Reston [Virginia] when the Space Station was headquartered there. I worked for Lockheed in developing proposals and a program-plan, budget-management plan to fly the large-format camera again. Later I worked with McDonnell Douglas on their section of the Space Station when they had prime responsibility for a certain part of it, which was later reorganized and the management responsibility was given to Boeing.

Subsequently, later to that, I worked with McDonnell Douglas in developing a quick reaction, skunk-works type single stage to orbit program where we actually in the period of eighteen months designed, built and flew a tenth-scale vehicle that we flew at White Sands [Test Facility, New Mexico] to demonstrate the capability of this particular vehicle, which was called the DC-X.

I think that just about concludes my activities before, during and after NASA.

Butler: You certainly did have quite an interesting and varied career.

Grimm: Then your next question here [referring to pre-interview notes], you said, who are the people that I worked with that made a significant impact on me professionally as well as personally. Well, I think, as I mentioned Warren North was the prime one. Slayton, to the degree that he more or less did let me do my own thing over there. I gained a lot of respect for Bob Gardiner.

Bob Gardiner originally said, "I don't care what you do, just don't cause problems and don't bother me." That was essentially what Max said. But later on he gave me some congratulatory salutations, and the same with Max. I think Max appreciated my management capability and my interest in doing things technically within NASA. He used to call me over and we'd have some interesting discussions.

So those are probably the three or four people who influenced my career with NASA.

Of course, as I said, this Ed Smith is the one that got me interested at NASA when I was with the FAA. He and I both had been at the FAA. Then this Captain Brickle [phonetic] at the time, who is now General Brickle, who was a very aggressive individual, more so than I am, I think.

So I think that covers that category.

Butler: Certainly a lot of interesting people to fit in with all the different areas in which you worked.

Grimm: My most important accomplishment in my career with NASA, per your question here [referring to notes], is I consider in the long haul that the Lunar Landing Training Vehicle project management program management and getting Neil trained in the nick of time so that they could have their FRR and then accomplish lunar landing. That's a significant accomplishment.

The second thing would probably have been developing the rendezvous techniques and procedures and training the crews on that.

The challenging milestone has to be the LLTV and the leading up to the lunar landing. I think that probably answers all of the questions unless you have some.

Butler: That covers most everything. I have one last one, to kind of tie things off.

Looking back to when you were first getting interested and first figuring out what your career was going to be, and you've mentioned the cards and the cereal boxes as a child and then working for the Air Force and so on, would you ever have imagined where your career would lead you?

Grimm: Never in a million years. I had no concept of where I was going to end up. That goes back to when I was seven or eight years old and I used to lay on my back looking at the stars at night and wondering what held the stars up and what was going on up there. I had a very early interest in things like that that I didn't understand, and I'd always hoped that I'd be able to get enough education to understand those things.

As things worked out, I was able to do that. As far as I'm concerned, it's been a very interesting career. As I said, up until the last three years of my time at NASA, my previous experience both in industry and at NASA were very challenging, very stimulating, and the last three years was sort of a drag in the management function.

Consulting was very interesting up to a point. But my wife liked it because she went with me everywhere.

Butler: That's a big perk.

Grimm: But consulting had limitations because you were there at the courtesy of whoever signed your contract. A lot of times you were wanting to contribute more than they asked of you. Sometimes I did contribute, and that didn't go over too well, when I saw the program going in directions other than what I thought it should. In most cases, I was proved right, so I feel vindicated in that respect.

The interesting thing about a lot of those consulting programs is that the reason why they get you as a consultant is that they are very naïve in those areas where you have expertise. Sometimes it's a very hard job to convince those people that they need to do certain things in these programs, because usually you're hired at a management level such as advising them on technical issues, management issues, developing management plans, program plans, systems engineering plans, systems integration plans, systems test plans, flight plans, if that were the case, things like that, new ways to do business quicker. I almost said Goldin's favorite words, quicker, better, cheaper.

But in fact, it was some of these programs, which is the now familiar term of skunk-works, you did things like that with very little supervision, lots of money, and you were able to get the job done and deliver a product that works. In the final analysis, sometimes, that's the key thing. So I enjoyed consulting for the most part and took my wife around, and she saw all the places that I had seen in the previous twenty-five, thirty years. In the process, we got to travel a lot. You can just wipe this last part off here. [Laughter]

Butler: Oh, that's all right. Well, you've certainly had quite a lot of important and interesting stories and contributions to the space program. I appreciate you sharing them with me and with the project.

Grimm: You're quite welcome. I think sometimes when I talk like this, it sounds like I'm tooting my own horn. As I said earlier, there's a lot of other people in most cases that were involved with these projects that I've discussed that made them happen and a lot of good people who probably may never get the recognition that they deserve.

But as a p.s. to this whole thing, that was one of the things that I always tried to impress on my people. I said, "You do good work for me, and I will do good things for you." I carried that philosophy through the whole time I was at NASA.

I think if you asked Duane Ross to look back at the records, I was criticized by not only people in my own directorate but by the Center's other directorates of getting too many awards for people in my organization, as a percentage of the awards given at any given time.

Butler: Not something you would criticize for, I wouldn't think.

Grimm: Some people stand with their hand out, other people go work for the handout. When I thought my people were deserving, I made sure that I worked hard to get them the recognition that I thought they deserved.

Butler: I'm sure they appreciated that.

Grimm: I hope they did. But that's not the reason; I did that for them because I thought they deserved it.

A lot of places you can spend your lifetime doing good work and never get recognized, and I think it falls to the management of an organization if he has people in his organization that do good work and do exceptional work that they should be recognized, not only from a salutatory standpoint but from a monetary standpoint.

I had people in that second organization that I had that had been in a certain grade level for years, and they were at a grade level that was below what they were authorized to be at. I questioned them about it. They said, "Well, it's because management doesn't think this position is very important, with respect to the rest of the organizations." That falls to the management of that organization as far as I'm concerned.

I said, "If you do the things that I ask of you, then I will see that you get a promotion." Lo and behold, every person that I said that to I got promoted and recognized with awards including the exceptional service medal, the distinguished service medal, commendations, monetary awards, and so forth.

One final word. The thing that I like the most is that one of my managers said, "The one thing that we could always [count on—you were hard on us, but always fair." I couldn't ask for better than that!]

Butler: Okay.

[End of Interview]

Return to JSC Oral History Website

Curator: JSC Web Team | Responsible NASA Official: Lynnette Madison | Updated 7/16/2010
Privacy Policy and Important Notices

Commentary From David R. Scott, Apollo 15 Commander on the 12/9/08 JSC Conference

September 2, 2009
To: Wayne Ottinger From: DRS
Subject: NOTES to accompany "DRS comments on the LLTV JSC Transcript"

The following Notes as identified by letter provide further explanation and discussion regarding DRS Comments on the LLTV JSC Transcript, Dec, 2009. For each Note, the Transcript topic (subject matter discussed) and [#] numbered DRS comments (in blue) are shown for reference.

These Notes as well as the corresponding comments are provided in the context of a near-term (years) human lunar landing program, and in particular the Constellation program. Should such a program become a longer-term endeavor (decades), then many of these comments may not be relevant. An example is discussed in "Note C" below, Automation & Robotics.

Note A. 2nd stage effect

neil new lltv safer (1 min 33 sec) (In Point at 1 hour 13 min 23 sec of NASA Tape 2)

[1]-- DRS comment: However, one must avoid the 2nd stage effect (see Note A)

The explanation of the term "second stage (system) effect" was brought to my attention in 2004 during discussions with the editors of the Apollo Flight Journal, primarily Frank O'Brien, who has a degree in computer science where the term was apparently first used in teaching. Frank's memo on the subject is summarized as follows:

"The Second System is one that, after an architect creates a wonderful, elegant *something* (software/building/ spacecraft/whatever), the *next* attempt is likely to be bloated with all the things that never made it into the first, more elegant iteration. The first time through, you are limited by resources and what you don't know, resulting is a design that is very focused on conservative, well defined requirements. "Mission creep" is less of a problem with First Systems, if only because the only "mission" is to get things working the first time."

You will note that this applies to any "system," not only computers. More can be found at the Apollo Flight Journal web site (http://history.nasa.gov/afj/), Another term for this is "Brooks Law" as is explained in http:// www.answers.com/topic/brooks-law, which also provides many links for more detailed discussions.

Note B. Mars capability

[1]... and the emerging tendency to include: (a) Mars landing capability (Note B);

Mars is a very, very difficult program, and very long-term in its planning, preparation and funding (which is of course no surprise). However, the "technology" to be used will be very different from the technology of today; e.g., as you might expect, the software programs of today will be stored somewhere in a dark corner of a forgotten warehouse (at best) – doubtful that this will be re-vitalized; the people who are fortunate enough to be assigned the Mars program will have their own technology

The Constellation program required Mars capability to be designed into the lunar capability. Attempting to combine lunar and Mars requirements in the design, development, and operations of a lander will compromise both, just as such expansion of capabilities has so many times in aircraft programs. Three things will happen to the lunar program if Mars capabilities are required in the lunar mission "architecture" (whenever it is formulated), the lunar program will experience: (1) increased cost; (2) increased schedule; and (3) increased risk. Let the Mars folks design and develop their machine when the time comes.

Note C. Automation & Robotics

[1]… and (b) automation and robotics (Note C).

Automatic (robotic, AI) capabilities are becoming quite advanced, they are challenging and they are fun to develop. But they are not necessary, or even desirable for a "manned" lunar landing -- they will introduce complex and additional failure modes during the mission as well as require the corresponding time and resources necessary for integration; test and checkout; software verification; procedures development (normal, malfunction, and emergency): C&W logic and signals; mission techniques; mission rules; simulation (such as launch abort sims due to time criticality); training; and real-time mission support,…among other factors (e.g., the age-old problem – if a red warning light flashes, what is at fault: the system or the indicator? And during the time-critical landing phase, the delay in assistance from MCC could cost the farm).

Automatic (robotic, AI) systems are best applied to two areas: (1) to relieve the human burden of repetitious, tedious, and boring activities; and (2) to allow humans to do something that could not be done without assistance from an "automatic" system (e.g., a precision landing on a runway during zero visibility conditions). Landing on the Moon is an entirely different matter –the surface of the Moon is irregular in all aspects and even with precision (e.g. virtual reality) planning and programming, it is unlikely that an automatic system will be able to "see" (interpret) the surface conditions as well as the eye. Automatic (robotic, AI) systems would be great for an unmanned landing, but they are unnecessary and even compromising for a human landing.

Airliners are excellent at automatic landings, and the ground-airborne systems are superb. Pilots are able to land an airliner in zero visibility -- that is, on Earth an automatic landing will be successful even when the human cannot see the target point. However, on the Moon, with current technology, the system cannot direct the vehicle to a suitable landing point; only the eyes and skills of a pilot can direct the lander to a suitable landing point. However, in the future, especially with better knowledge of the landing point (higher resolution photos, better gravity models, etc.), an automatic landing could very well be as effective as an Earth auto land – but this advanced capability will obviously require time and money.

Note D. Need for an LLTV

gene risk here not on the moon (0 min 42 sec) (In Point at 1 hour 15 min 09 sec of NASA Tape 1, Disc 2)

[2]-- DRS comment; The LLTV risk environment on Earth is far less than the lunar lander (LM) risk environment at the Moon. In the Earth environment, the situation is under better control, and it is shorter term with a simpler vehicle. And at the Moon, both the mission and the crew are at risk – a risk that can be reduced

by having a qualified and proficient pilot at the controls. Only an LLTV-type vehicle that very closely replicates the LM can provide the qualification and capability necessary for a successful lunar landing (Note D).

The experience. Landing on the Moon is a brief, and very unforgiving, experience. Many factors are involved – each and every factor must be considered and evaluated continuously. These factors include vehicle dynamics and motion, control and handling qualities (including response time), landing point selection, time available, control systems operation, computer operation, and the operations of all of the other many vehicle systems. Therefore, it is essential that as many of these factors be integrated into the pilot's training and proficiency as possible.

What's the task of a free-flying Lunar Lander Training Vehicle? (A) To place the pilot in the control loop as an active (direct) and feedback element. (B) To condition (train) the pilot in the highly dynamic and short time-constant flight operations. (C) To enable the pilot to readily and comfortably enter an effective performance "zone" during landing operations (see Note J).

Pilot qualification and proficiency. Not only must the pilot demonstrate qualification in the LLTV, but the pilot must also demonstrate a high degree of proficiency in the LLTV. Otherwise, the pilot will not be suitable for a lunar landing (see also Note L.) As the old adage reminds us:

Aviation itself is not inherently dangerous.
But like the sea, it is terribly unforgiving
of any carelessness, incapacity, or neglect.

And the STA, albeit a very good trainer should not be used for analysis or comparison – the capabilities, objectives, and training benefits of the STA and the LLTV are entirely different. As examples, Shuttle approach and landing can be simulated in various aircraft (systems and procedures in other simulators); whereas the LLTV training for lunar approach and landing is absolutely unique (no other trainer, including Langley, could be used for this objective). See Note P.

Note E. LLTV Handling and Control

Gene do all you can (1 min 15 sec) (In Point at 1 hour 05 min 23 sec of NASA Tape 2)

[3]-- DRS comment: Agree. The LLTV configuration and controllability (handing qualities) are absolutely unique and tailored to replicate the same on the LM (Note E). It is unlikely any other flight vehicle, modified or not (currently known to the industry) can perform the same function.

The motion of a lunar lander is absolutely unique. In particular, the 3-axis horizontal and vertical velocities are strongly and instantly coupled as functions of engine thrust level and vehicle attitude (R, P, and Y). that is, the 3-axis translation depends on the relative vectors, which can be controlled by the throttle as well as the attitude (pointing) of the vehicle. Therefore, only a free-flight LLTV-type vehicle can be used for realistic and efficient simulation. These multi-variable operations cannot be adequately simulated in a fixed-base or moving-base simulator. Further, the LLTV-type free-flight motion cannot be simulated by a helicopter or hovercraft (either of which can however simulate the flight along the landing trajectory and/or directional path).

"Aircraft" do not have the dynamic response or handling qualities of a lunar lander, and they are also subject to aerodynamic forces that would be difficult to filter or cancel – thus they are not practical for lunar landing simulations. However, helicopters are quite useful in becoming familiar with steep descents and are a valuable precursor to an LLTV type vehicle.

Note F. Flight Vehicle Experience

john young no lltv, crashes (0 min 13 sec) (In Point at 09 min 48 sec of NASA Tape 1, Disc 2)

[4]-- DRS comment. The actual LLTV flight record must be analyzed before such a comment is made. As an analogy, not unlike fighter squadrons in the 50's, upon the introduction of a new flying machine, problems occur, often serious problems. Thereafter, based on experience and maturity of the flight vehicle, the pilots, and maintenance, the problems diminish significantly (Note F). …

Historically, upon introduction, new flying machines go through a transition period of learning and experience which is often characterized by "accidents." In recent times, one of the most advanced flight vehicles ever designed, the F-22, has unfortunately experienced three accidents. One would now expect significantly improved F-22 operations in the future.

During earlier times, and as an illustration based on my personal experience, some years ago I joined a fighter squadron that had an exceptional safety record, with highly experienced pilots, ground crews, and maintenance – over four years of F-86 operations, some 40,000 hours without an accident. We then transitioned into the F-100 – during the next 14 months this same squadron experienced 8 accidents (major and minor); thereafter the operations smoothed out with only 3 accidents during the next 4 years (2 due to ground radar flight control) – a maturity record not unlike the LLTV.

Note G. LLTV Experience

[4] …..The LLTV is typical – after 4 years of introduction with 3 losses, during the following about 3 years (1 year overlap), LLTV #3 flew 286 flights without loss, through the end of the program (Note G). Thus the comment that "we crashed three out of four" is not a proper assessment, especially for a program that matured and concluded at a high point of success.

Note H. Failure Detection

schmitt manual ldgs (0 min 21 sec) (In Point at 0 hour 58 min 10 sec of NASA Tape 1, Disc 2)

[8] -- DRS comment. "Manual" landing actually means "pilot-in-the-loop." We called these manual landings because the pilot was controlling the vehicle manually (by hand) – but in actuality, the hand-controller inputs were going through the computer, which itself could have landed the LM automatically (albeit never used). Being in the loop provides more than just a good feeling (which it definitely does) – being physically in the control loop has several operational benefits: (a) the pilot is tuned, or on top of the situation, -- if a failure occurs during an auto land, the pilot must recognize (accept) the failure and instantly insert himself into the control loop; which requires a finite amount of time; and (b) being in the loop at the time of a failure removes the ambiguity of determining whether or not an indication of a failure is actually a failure of the system or a failure of the indicator (see Note H).

"On Tuesday afternoon, after NASA fueled the shuttle's huge external tank, one instrument showed that the fueling valve failed to close, while other indicators told launch controllers that the valve closed fine. With that uncertainty, the launch team scrubbed Wednesday morning's launch attempt and decided to try for shortly after midnight Friday. Now NASA is looking at 23 hours and 37 minutes after that, 11:59 p.m. Friday" ("NASA scrubs tonight's launch; next opportunity is Friday night." Orlando Sentinel, August 27, 2009).

Imagine attempting to troubleshoot something like that during the final 2 minutes of a lunar descent and landing…!!

Note I. Vehicle Motion

I knew it was – been in the tv (0 min 10 sec) (In Point at 1 hour 01min 08 sec of NASA Tape 1, Disc 2)

[13] - DRS comment. Another factor is the motion that the pilot feels, especially while maneuvering briskly – and correspondingly a motion that the LMP is not familiar with. That is, unexpected physical motions and forces are distracting, and the pilot in particular must be inherently tuned to this motion (see Note I).

To illustrate this point, it is useful to review the onboard transcript and comments from Pete Conrad and Alan Bean as recorded in the Apollo Lunar Surface Journal.

110:31:06 Bean: Oh! Look at that crater; right where it's supposed to be! Hey; you're beautiful. Ten percent (fuel remaining). 257 feet, coming down at 5; 240 coming down at 5. Hey, you're really maneuvering around.

110:34:08 Carr: Roger, Pete.

{I asked Pete about the value of this training.]

{Conrad - "I think everybody agreed that the LLTV was very essential to a successful landing. One of the problems which we were talking about earlier at lunch is that you have to realize that the visual on the simulator was very bad. We had a plaster-of-paris lunar surface (called the L&A) and a B&W television camera that (flew to it). So you're looking at a flat, no-depth boob tube - a television - in the window. So, the last five hundred feet, if you were watching that out the window in the simulator, it wasn't any good. It just didn't really resemble the real world."]

{Conrad - "It was the best they had at the time. We didn't have a moving-base simulator, either. (In more recent times, Shuttle crews have been able to train in simulators that move in response to crew inputs and, thereby, give much more realistic simulations.) So, the LLTV was critical, to get a real feel. And the reason Al made the comment about maneuvering is that the LMP's didn't fly the LLTV."]

{Conrad - "Al hadn't flown one, and that's why he made the remark when I started really maneuvering the thing around. Because you had big attitude changes up there, because you're in a low gravity field. He had seen that kind of a maneuver, probably, inside the simulator. But that virtual image display and the fixed base didn't really give you any feel for it. So, the first time Al really experienced that was at the Moon. And I just passed it off 'Yeah, I'm busy doing what I was doing.'!"]

{Bean - "We were all flying helicopters and you didn't maneuver a helicopter any where near like that. Up there, you really had to move the LM to maneuver it. So Pete got used to it and I was thinking helicopter kind of stuff. So, when you (Pete) suddenly maneuvered much more than a helicopter, it caught me by surprise. But to you, well, that's the way you do it. I think it's because, on Earth, you're supporting the weight with a certain amount of thrust.

So, let's say you've got to knock off ten foot per second forward. You pitch up to a certain angle to do that and you get used to that kind of maneuver. You go on to the Moon; you've got one sixth the thrust to hold this same mass up and ten feet per second forward with that mass is the same as it was on Earth. But, in order to stop it (from moving forward) with one-sixth the thrust, you're going to have to pitch up a lot harder. So I think it's just strictly the fact that you're operating with less thrust than a helicopter for the same weight and the same momentum. So, in order to use it, you've got to get that thrust vector up higher faster or you're just never going to slow down the translations, or get one going and then stop it. When you think about it, it makes sense. But, at the time, it just seemed like 'God, what's he doing?' It felt to me like you were pitched too far (back), you know. And you probably were doing quite a bit because you've got to get it (pitched) up there to get the little ol' thrust vector to work."]

{Conrad - "That's right, you have to move it more to get the maneuver. So it looks really bad to you, although nothing serious is happening."]

{Bean - "It looked normal to you!"]

This shows that without LLTV training, the pilot will be surprised by the motion during actual landing, and of course distracted as well as questioning whether or not the LM is behaving properly an/or whether or not there is a failure in the control system.

This also introduces in the case for fixed-base vs. motion-based simulators. Motion in space is benign (except tumbles); motion during landing is dynamic and time of flight is extremely limited. An LLTV is essential for the dynamic elements of a landing simulation – a fixed-base simulator is fine for procedures development and systems training. Therefore, the combination of an LLTV and corresponding fixed-base simulators is optimum; both from training and cost perspectives. Learn to fly the LLTV and learn procedures and systems in a fixed-base simulator (no need for a motion base).

Note J. Zoning

gene neil training reactions same (0 min 36 sec) (In Point at 1 hour 33 min 00 sec of NASA Tape 2)

[14] - DRS comment. This could be considered part of the "zoning" process (Note J) – we all learned to completely focus on the intense and demanding task of landing a vehicle on the moon. We were all trained into essentially the same performance box. As an analogy, we all (most of us) land airplanes in approximately the same manner.

"Zoning" (also known as "Flow") can be explained as "the mental state of operation in which the person is fully immersed in what he or she is doing, characterized by a feeling of energized focus, full involvement, and success in the process of the activity." Initially proposed by psychologist Mihály Csíkszentmihályi, the concept has been widely referenced across a variety of fields. Colloquial terms for this or similar mental states include: on the ball, in the zone, or in the groove" (http://en.wikipedia.org/wiki/Flow).

This was not a familiar term during Apollo, but this is what we did – during descent and landing; and because of LLTV experience, we actually entered what is now known as "The Zone"-- for lunar landing. The term is now widely used in sports psychology and clearly defines our "psychology" in both training and actual operations. To better understand the zoning concept as it relates to the LLTV and a lunar landing mission, review one of several published explanations of zoning; e.g.., "Entering 'The Zone': A Guide for Coaches" (http://www.thesportjournal. org/VOL2NO3/COSTAS.HTM). Also see on the internet: "Using Sports Psychology to Achieve a Zone Focus."

As a final comment on this, see the lengthy discussion of the Apollo 15 Landing at 102:42:48 in the Apollo Lunar Surface Journal (http://www.hq.nasa.gov/alsj/frame.html), part of an interview with Eric Jones during 1992-1993 -- note that even then, I felt that I was on "automatic."

"The LLTV landings were manual landings, and the LLTV was a great trainer. I mean, boy, am I glad we had that, because it gave me confidence that I knew what I was doing on the Moon, and I didn't have to think about things. I didn't have to consciously program myself to do things. I was automatic. So, my feeling was, if you can land the LLTV, you can land a LM."

Note K. 2nd (Safety) Pilot

Reset Button & Altair Gimbal RV Sim (5 min 17 sec) (In Point at 1 hour 40 min 15 sec of NASA Tape 2)

[25] - DRS comment. The LLTV does put you on the line and it forces you the think the lunar landing profile (Note J) – it does not have a reset button such as a simulator, but it does have a breakout capability to exit the lunar sim and then land in a much easier and safer mode. A safety pilot actually increases risk (see Note K).

I strongly favor the solo LLTV configuration (pilot only) for the following reasons, among others:

1. What would the 2nd (or "safety") pilot do?

   - Call out corrections? (and interrupt the pilot – the Flight Director can communicate systems problems).
   - Make comments during an intense maneuver? (and slow the pilot's thought process – the pilot should not be distracted by 2nd pilot opinions)
   - Act as the LMP? (The LMP gets his training in the LMS; and the PILOT must be able to land without LMP communications anyway)
   - Grab the controls if he does not like the situation?
   - Get a thrill; be frightened.

2. What controls and displays would be added for the 2nd seat? What overrides? All of which would need to be integrated into the total system and result in additional failure modes, more complex mission rules, and more complex flight operations.

3. Adding a 2nd place will increase cost, schedule, and most importantly risk – for no recognizable return (that is, it is unlikely the 2nd seat would add anything meaningful to the training, and indeed subtract from it).

4. No time to do anything that would contribute to safety (but could detract from safety).

5. Distracting to the pilot.

6. Might provide reliance by the pilot in certain situations where the pilot should make the decisions rather than rely on a safety pilot – thus perhaps providing a false sense of security

7. Communications might conflict with and confuse the comments from the Flight Director (who would have much more data)

8. Pilot decisions must be based on the task at hand; and not based on having a passenger for which the pilot is ultimately responsible.

9. How many of the 2nd seat people would have survived the three losses of the LLTV during Apollo?

10. Certainly not going to give the 2nd seat the trigger to eject.

Remember that the LLTV pilots are not (should not be) beginners; they should already be comfortable in solo checkouts of new and/or high performance flight vehicles (e.g., grads of a recognized TPS).

Note L. Management Responsibility

scorch $ should not drive (0 min 28 sec) (In Point at 1 hour 13 min 24 sec of NASA Tape 1.Disc 2)

[27] - DRS comment. But money is a reality and must be a factor in management's analysis. However, management's responsibility in committing the crew to a landing must also be weighed in the tradeoffs of LLTV cost and risk (Note L – very important).

The responsibility of senior management is to ensure the highest probability of success of the mission coupled with minimum risk of loss. The LLTV-type vehicle (LLTV) itself contributes to both. But the pilot of the Lunar Lander (LM, Altair, etc.) must be proficient in the LLTV; that is he/she must have demonstrated – repeatedly -- very high-quality flying capabilities and flight-management skills. To send somebody to land on the Moon (planet, etc.) who has not proven him/her-self in an LLTV-type vehicle would be irresponsible – that is, without demonstrated capability in the LLTV, a non-qualified pilot would lower the probably of success and increase the overall risk of the mission.

To quote Pete Conrad: "We are banking our whole program on a fellow not making a mistake on his first landing." Digital Apollo, David A. Mindell, MIT Press 2008 (p 181).

Note M. Time Compression

Neil 11 gene psycho (4 min 30 secs) (In Point at 0 hour 44 min 07 sec of NASA Tape 1, Disc 2)

[28] - DRS comment. The fuel limits were a very valuable contribution to the overall preparation for the LM landing. This limit forced early decisions on the approach and touchdown, which if anticipated gave us time to change the touchdown point during the final gallons of fuel – something to learn based on the actual Apollo 11 experience. This additional limit also helped compensate for time-compression during the actual mission (Note M).

As quoted from Ken Szalai (2-21-2008) in the Go For Lunar Landing Conference charts, March 4th and 5th, 2008: "The lifting body pilots were unanimous in reporting that, once in flight, the events of the mission always seemed to progress more rapidly than they had in the simulator. As a result, engineers and pilots experimented with speeding up the simulation's integration rates, or making the apparent time progress faster. They found that the events in actual flight seemed to occur at about the same rate as they had in the simulator once that simulation time was adjusted so that 40 simulator seconds was equal to about 60 "real" seconds."

This ratio of about 7 to 1 was just about the same as I believe most of us experienced during Apollo; that is, it seems that the time available in flight is about 70 % of the time used during simulations.

One of the common phrases was "get ahead, and stay ahead." NASA never implemented this simulation philosophy, and I don't know why. But the LLTV certainly taught us this valuable lesson.

During Apollo 15, I went to manual at about 400 feet, and landed 81 seconds later (55 seconds sooner than the propellant budget provided). Observing the DFRC experience in time compression during a flight, whereby 60 actual seconds seems like only 40 seconds of available time for operations, my 81 actual seconds probably seemed like 54 seconds (it certainly went quickly). A typical LLTV profile from 400 feet to touchdown takes about 69 seconds – great training for the 54 flight seconds that I experienced. Thus, the available time, and time usage, during my lunar landing felt very similar to my experiences in the LLTV.

And this was a most important lesson, because (to again quote from Ken Szalai); "…the consequences of fuel exhaustion were nearly the same for the LLTV mission as for the LM landing." But of course there was much more on the line during the lunar landing; even more reason to be well schooled in this art.

Note N. Negative Training

Several items have been identified as negative training in this discussion; among these the following should be emphasized.

Note N-1. Propellant

Neil 11 gene psycho (4 min 30 secs) (In Point at 0 hour 44 min 07 sec of NASA Tape 1, Disc 2)

[28] … Further, having more fuel (time) than actually available during the lunar landing would be negative training, it would preclude the pilot from integrating the time factor into his thinking (Note N).

Time to landing is a direct function of propellant remaining. The LLTV flight profile approximates very closely the time remaining for the lunar lander during a near-identical flight profile. The time remaining during descent (or propellant available) must be integrated into the pilot's flight profile and decision-making. Therefore, providing an excess amount of propellant during an LLTV flight would provide negative training in that the pilot might become complacent regarding the amount of time remaining before a commitment must be made to land or abort (regardless of warning lights or any other indications).

Note N-2. Systems

Complexity of off-nom tng vs skill tng (3 min 39 secs) (In Point at 1 hour 07 min 14 sec of NASA Tape 1, Disk 2)

[30] - DRS comment. The purpose of the LLTV is to teach the pilot how to fly, not how to analyze systems that are different from those of the LM – for the LLTV, that's the job of the LLTV flight control team (CRM) (and for the LM, that's the job of the LMP and MCC). Therefore, absolutely no need or purpose to insert intentional systems failures or simulated failures in the LLTV for training; those would be another aspect of negative training (Note N).

The inclusion of LLTV systems failures during an LLTV flight, would result in the following negative factors (among others).

1. Need for pilot to become proficient in systems additional to and different from the LM (and CSM) – requires more training, documentation, and coordination with the CRM.

2. Distracting from the primary purpose of the LLTV – to teach the pilot how to fly.

3. Increases time, effort and complexity of pilot training.

4. Increases overall cost, schedule and risk of the LLTV.

5. Different systems in a similar vehicle (LLTV) could become a confusion factor in the pilot's operations of the actual LM.

Note O. CDR Window Display

neil rate of descent cmd (1 min 37 sec) (In Point at 1 hour 01 min 56 sec of NASA Tape 1, Disc 2)

[35] - DRS comment. The ROD was great; but still, becoming proficient in the LLTV "manual" landing is most important – for both the potential failure of the ROD as well as to appreciate the ROD function. Another recommendation to reduce the pilot's load in the cockpit is to add a pipper to the LPD grid; that is, a "dot" on the CDR's window that moves along the grid according to the two-digit numerical computer projections of the landing point. This would correspond to (or substitute for) the DSKY numbers being read by the LMP. Any other CDR window display would be superfluous and detract from the visual and mental requirement of selecting and flying to a specific touchdown point (Note O).

The verbal transfer of information was not cumbersome and in fact optimized the landing technique – that is, "we" landed on the Moon – the machine (LM), the CDR (eyes outside), the LMP (eyes inside), and MCC (many eyes). Therefore, the LPD as a representative HUD would be fine. Otherwise, don't block the CDR's view or distract him with information that can be provided through his ears by the LMP and MCC.

As an example, on A-15, we found that the optimum technique was for the CDR to focus out the window and the LMP to focus inside the cockpit. The CDR maximized the landing point information through his eyes while absorbing the LM gages and systems through his ears (LMP and MCC); thus optimizing all sensors available. The verbal transfer of information from the LMP to the CDR was a most effective technique to optimize the use of all of the information available – that is, with proper training and coordination, each of the LMP and CDR could focus on assigned tasks without dilution, confusion, or subtracting from the maximum use of available information.

The lunar landing as we implemented it on A-15 was not unlike landing in formation during bad weather – one pilot has his eyes in the cockpit scanning all of the information available and the other has his eyes outside searching for the runway – when the outside eyes see the runway, "contact" is called and the lead comes out of the cockpit and begins the flare to touchdown. Prior to that, the lead is completely inside the cockpit "on the gages." Both hear ground control, e.g., GCA, and the lead reacts to the information provided.

A HUD on the LM that duplicates most of the data being read to the commander (CDR) would require the CDR to scan, refocus and interpret the display. Also, situational awareness, especially when landing on the lunar surface, includes "terrain awareness" as well. With the field of view cluttered up with pretty but marginally useful graphics, some of that awareness would be lost. Further, any failure in the display itself would be another undesirable distraction.

However, adding a pipper to the LPD grid would be helpful -- that is, superimposing a "dot" on the LPD grid that moves along the grid according to the two-digit numerical computer projections of the landing point. This would correspond to (or substitute for) the DSKY numbers being read by the LMP.

In summary, adding an LPD pipper to the CDR's window would be fine. Otherwise, don't block the CDR's view or distract him with information that can be provided through his ears by the LMP and MCC.

Note P. STA Comparison

gene roger eye level (1 min 10 sec) (In Point at 0 hour 0 min 0 sec of NASA Tape 2)

[20] - DRS comment See [17] above. However, regarding the STA comment, the physical configuration of the LLTV cockpit, especially controls and displays, should be exactly the same (or as close as possible) to that of the lunar lander. A blindfold cockpit check of all controls and displays is mandatory before proceeding to "checkout" in the lunar lander – standard aeroplane practice. Also, in general, caution on comparisons or analogies with other trainers, especially the STA (being a recent example of a unique trainer). See [23] below, and Note P.

gene Mash free flt lltv (1 min 47 sec) (In Point at 1 hour 15 min 01 sec of NASA Tape 2)

[23] - DRS comment. The lunar landing is so unique, that an LLTV type vehicle is essential. Conversely, for the Shuttle approach and landing, if the STA were not available, training in an F-104 or similar low L/D aircraft would be adequate, the profile and judgment requirements are essentially identical, and control responses are similar. No known vehicle, other than the LLTV, can replicate the control responses of the lunar lander (LM) (Note P)>

The following comments are intended to illustrate the different characteristics of an STA and an LLTV as they relate to the design and operations philosophy of training a pilot for a landing. The comments have been drafted primarily for non-pilots, engineers or managers – most of the differences between the LLTV and an STA are probably obvious to those familiar with aircraft flight operations. However, these comments are also important for those skilled in aircraft (or helicopter) flight to understand the differences introduced by lunar-landing flight. In general, the STA and LLTV are obviously completely different flying machines with completely different flight profiles. Extrapolation from the SVA design and operations philosophy to the LLTV will most likely result in, among other problems, negative design considerations and negative training.

As background, I was very fortunate during my flying career to have had the opportunity to fly many "low L/D" approaches (and touchdowns). I was also fortunate to have had the opportunity to fly the LLTV. These are entirely different machines flying entirely different profiles. Many of the following comments may be obvious to those skilled in the art, but a review might be useful to assist a focus on the current situation. Also, the following comments are not meant to compare the STA and the LLTV directly since they fly in different regimes, it is meant to compare the design philosophy of simulating an aerodynamic vehicle that flies in a 1-G environment with simulating a different vehicle – under aerodynamic and 1-G conditions -- that flies in a 1/6 G vacuum environment.

1. Flight profiles

   (a) The low L/D approach profile (STA, Shuttle, X-15, X-24, F-104, etc.) is an energy management technique based on proper positioning at a high gate, followed thereafter by careful attention to altitude, airspeed and approach angle. Only two degrees of freedom are required, roll and pitch; the vehicle then flies as a function of aerodynamic forces and gravity; propulsion is not involved.

   (b) Conversely, the LLTV approach profile (manual) requires the same energy management skills, however, the profile is a function of roll, pitch, and engine thrust (throttle position) -- the vehicle then flies as a function of engine thrust and gravity. At the same time, the LLTV systems must in fact eliminate (compensate for) aerodynamic forces.

2. Vehicle control. The Shuttle (and STA) requires the pilot to use: (a) the rotational hand controller (RHC). The lunar lander (and LLTV) requires the pilot to use: (a) the RHC; (b) the throttle; and (c) the coupled combination of the RHC and the throttle. Therefore, the LLTV requires much more challenging pilot control and coordination, and correspondingly a more difficult training task.

3. Response time. The motion of a lunar lander is absolutely unique. In particular, the horizontal and vertical velocities are strongly coupled with -- and an instant function of – vehicle attitude; that is, the 3-axis translation depends on the relative vectors – these are controlled by the throttle as well as the attitude of the vehicle. The response time to "control" inputs is almost instantaneous -- not so with the STA, helicopter, hovercraft or ay other vehicle that relies on aerodynamic forces for motion, and that can be flown with relatively long time constants.

   In comparison, things happen much more quickly in a lunar landing approach than in a low L/D approach – therefore, during an LLTV approach the pilot must anticipate necessary corrections much sooner than in a low L/D approach. In other words, and not to detract from the challenge of a low L/D landing, a low L/D landing is much less challenging than a lunar landing.

4. Propellant. Other than a fixed throttle setting, the STA requires no additional propellant to maintain its flight profile. The LLTV requires engine modulation, close monitoring and conservation of propellant, especially as the approach relates to selecting a desired touchdown point. Thus, the STA: no throttling required. The LLTV: the throttle, and the use of propellant, controls: (a) the remaining flight time, (b) the descent rate, and (c) the horizontal velocity (forward, aft, left, right).

5. Flight profile and time to touchdown – (considering the differences beginning at about 300 feet). The STA requires only a careful initiation of the flare and holding the pitch as the aircraft touches down. Conversely, this region is one of the most active periods of an LLTV flight in that roll, pitch, and throttle movements (coupled) may be quite rapid in order to find a suitable touchdown point.

6. Touchdown point. Most importantly, the STA makes its approach to a specific touchdown point on well-marked, smooth, level, specific-width surface (runway). The lunar lander makes its approach to a surface that not only has no specific orientation markings, but a surface that is distorted by craters, rocks, shadows, slopes, mounds, and other irregular features – and final touchdown is often IFR due to the dust. Yes, in training, the LLTV works to a runway, but the pilot has the opportunity to change the touchdown point at the last minute to simulate undesirable lunar surface conditions, whereas the STA touchdown point is fixed; that is, there is no reason for the STA pilot to change the planned touchdown point – however, as we have learned, the lunar lander pilot must often change the touchdown point at the last moment.

DRS

The Collier Trophy for Four X-15 Pilots --- My X-15 Experience 4/1960 -- 12/1962
Solving the Orbital Rendezvous Challenge
Joe Fletcher's Story of Landing on the North Pole
Two photos & One Letter from late a discovery

The Collier Trophy --- My X-15 Experience 4/1960-- 12/1962

In a White House ceremony on July 18, 1961, President John F. Kennedy presented
the Collier Trophy to X-15 pilot Major Robert M. White (shown standing next
to the Trophy). Also receiving the award were Commander Forrest S. Petersen
and Dr. Joseph A. Walker (not pictured). (NASA photo no. 62-X-15-19).

Dr. Joseph A. Walker stands beside the 1961 Collier Trophy, awarded to him and the other X-15 Pilots by President John F. Kennedy. (NASA photo no. 620X-20).

Joe Walker WWII DFC P-38s
As the X-15 Flight Ops Propulsion Engineer & Later as the LLRV Project
Engineer working with Joe as he made the First Flight of the LLRV

Joseph A. Walker

Joseph A. Walker was a Chief Research Pilot at the NASA Flight Research Center, currently Armstrong Flight Research Center, during the mid-1960s. He joined the joined the National Advisory Committee for Aeronautics (NACA) at the Aircraft Engine Research Laboratory (today the NASA Glenn Research Center) in Cleveland, Ohio, in March 1945 as physicist and research pilot heavily involved with icing research. In 1951, he transferred to the High Speed Flight Research Station at Edwards Air Force Base, where he spent the rest of his career.

At Armstrong, he served as project pilot at the Edwards flight research facility on such pioneering research projects as the D-558-I, D-558-II, X-1, X-3, X-4, X-5, and the X-15. He also flew programs involving the F-100, F-101, F-102, F-104, and the B-47.

Walker made the first NASA X-15 flight on March 25, 1960. He flew the research aircraft 24 times and achieved its fastest speed and highest altitude. He attained a speed of 4,104 mph (Mach 5.92) during a flight on June 27, 1962, and reached an altitude of 354,300 feet on August 22, 1963 (his last X-15 flight).

He was the first man to pilot the Lunar Landing Research Vehicle (LLRV) that was used to develop piloting and operational techniques for lunar landings.

Walker graduated from Washington and Jefferson College in 1942, with a B.A. degree in physics. During World War II he flew P-38 fighters for the Air Force, earning the Distinguished Flying Cross and the Air Medal with Seven Oak Clusters.

Walker was the recipient of many awards during his 21 years as a research pilot. These include the 1961 Robert J. Collier Trophy, 1961 Harmon International Trophy for Aviators, the 1961 Kincheloe Award, and 1961 Octave Chanute Award. He received an honorary Doctor of Aeronautical Sciences degree from his alma mater in June of 1962. Walker was named Pilot of the Year in 1963 by the National Pilots Association.

He was a charter member of the Society of Experimental Test Pilots, and one of the first to be designated a Fellow. He was fatally injured on June 8, 1966, in a mid-air collision between an F-104 he was piloting and the XB-70.

**Commander Forrest S. Petersen, Later Retired as Commander
of the Naval Air Systems Command in i980
I Met With Him Twice In The Pentagon in 1972**

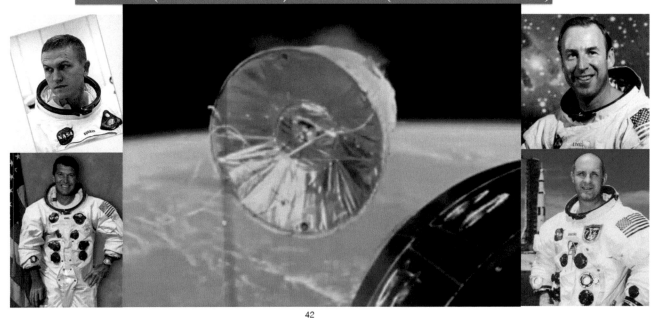

The First Successful Rendezvous of Two Spacecraft in History
Gemini 7 (Borman & Lovell) as the Target and
Gemini 6a (Schirra & Stafford) as the Hunter (Just under Six Minutes)

42

The First Rendezvous in Space

The First Rendezvous in Space History

The Gemini 4 mission's first attempt at rendezvous by Jim McDivitt and Ed White on June 3, 1965, failed due to a lack of understanding by the mission planners of the orbital mechanics required to accomplish success. This was a critical requirement for the Apollo program. It was mandatory that these maneuvers be safely accomplished to fulfill the overall mission of getting to the moon. There is a remarkable story to tell how the challenge was overcome, and success achieved on December 15 1965, just six and one-half months later. The above picture of the Gemini 6a crew, Wally Schirra & Tom Stafford on the right, targeting Gemini 7 in orbit with Frank Borman and Jim Lovell on the left aboard during their 14 day mission. Now, for the rest of the story.

Dean Grimm had been managing Wayne's LLRV & LLTV's programs for the NASA Manned Spacecraft Center or MSC, now the Johnson Spacecraft Center. He temporarily exited that program and was assigned primary responsibility for developing new orbital mechanics to correct the failure of the first rendezvous attempt made by the Gemini 4. Dean told Wayne this story soon before his death in April 2014. Dean said that NASA MSC at Houston had just installed a $40 million mission simulator and he requested that they run the rendezvous simulation backwards so he could identify maneuver errors and develop corrective actions. MSC computer experts, said *"no, it could not be done."*

So, Dean contacted his classmate from the University of Kansas, who was running the McDonnell Douglas Gemini mission simulator at the St. Louis facility. Dean got a positive response, and went through NASA headquarters to get permission to take the two astronauts assigned, Wally Schirra & Tom Stafford, to the next mission apollo 6a to work with him. on Gemini 6a,, They used the St. Louis simulator to solve the problem. Dean got the go-ahead and about five months later they presented their rendezvous solution at NASA headquarters. NASA's Red Team, Boeing, recommended their solution would not work, and that NASA give

Boeing the task to develop new rendezvous procedures. Dean immediately announced that he would resign from NASA, should they accept Boeing's recommendation. Schirra also said he would resign from the mission, unless NASA accepted their solution. NASA accepted their solution from the Gemini simulator in St. Louis, and one month later the Gemini 6a mission succeeded. Subsequently Dean retuned to our LLRV and LLTV program.

<u>From my 1970s story</u> From a Fletcher collogue at the Ohio State University
Excerpts From Joe Fletcher Oral Interview -- First to Land on North Pole

BS: When did you go to North Pole?

JF: Well, I wanted to do a bunch of things and I . . . once we got started, I wanted to keep the ski C-47 at T-3 and get a little fuel carried in and several things I wanted to do. One, I wanted to visit and possibly recover the R4D that Ed Ward had left on the ice. Do you remember that?

BS: No.

JF: In Ski Jump II, it didn't go . . . Ski Jump I went very smoothly.

On Ski Jump II, they had the P2V and they were having trouble with the engines and various things, and the upshot was that they hit a bump in the ice and busted up a ski and they abandoned the R4D out at one of the stations.

BS: Which station?

JF: I don't know. Well, one of the hydrographic stations. BS: I see, on the sea ice.

JF: On the sea ice. That's right. And, of course, that had not been seen or that was written off. But from talking with Ed Ward, it appeared the airplane was in pretty good shape and if you could get it off, it would be probably otherwise OK. So, I had hopes of visiting the site and if it were feasible to do, maybe even recovering the aircraft. That was one thing. Another thing I wanted to do was to go over and land on T-2 and be able to explore it and look at the conditions and so on. I wanted to land on the coast of Ellesmere because that's where we thought the ice would come from, originally. And so you really have to make a visit to get anywhere with that. So, those were the things that I wanted to do. And so, I asked the Air Command for the use of the C-47 for 30 days. Well, as I say, the staff was all against it because this was an invitation to disaster in their minds. But General Old, by that time, was rather sympathetic to my wishes and he wanted to say yes, but he was being told by his staff to say no. So, the upshot was that after we were set up there and this was, by this time, we'd gone in the middle of March and by this time it was about the middle of April. The C-47 was scheduled in again and I wanted to be able to keep it. General Old had his own problems which I knew nothing about at that time, but anyway, we had been in communication by radio and I had insisted that I wanted the airplane and he had said no, and I had reiterated the request. Anyway, when it came in he came in with his instructions, and he said, "OK, you can keep the airplane for a month on one condition and you can do anything you want to, or in your judgment you want to do during that time. But there's one condition. The first landing that you make anywhere else except T-3 has got to be at the North Pole." And he was very explicit. And he meant it. So, that's what we did. And I didn't understand until later why he did it.

(350)

And the reason, it turned out - not the whole reason, but one of the reasons, was that he was also sympathetic to another fellow who I'd never met. What the hell's his name? I'll think of it in a moment. But this was a World War II fighter pilot who had a very good war record. He had, I forgot how many airplanes he'd shot down, but he was an Ace. But, he was a crazy guy. And he had a whole bunch of accidents of various kinds. He'd been court martialed a couple of times and gotten in trouble, but the general liked him because he was an Ace with a good record. And so he had bailed him out of a couple of embarrassing things. Well, he had gotten into another one. He had slugged his co-pilot. They were on some kind of a mission over around Nome somewhere and his co-pilot had objected to a couple of his shenanigans and so he had slugged him. And it had been quite an episode. So, there were charges pending. Bill Benedict. That was his name. So, General Old was trying to think how to get Bill Benedict out of trouble and I didn't know any of this at the time. But anyway, he had picked Bill to take the C-47 on the mission up to T-3 and the idea was to get Bill involved in something which would give General Old an excuse for bailing him out of this latest mess. And so, one fine day, Bill Benedict showed up. This was, as I say, it was some time later. About the middle of April by then, and the C-47 there and instead of Lou Erhardt, it was Bill Benedict that was the pilot and I had a note from General Old giving me these instructions, you know. This is the airplane you can keep.

BS: Here's your pilot.

JF: And, yeah. And the first landing has got to be at the North Pole and after that, you can do your own thing. Well, I followed the instructions. That's all. We waited about 10 days or so to make sure that the weather was just right. And then we made the trip up to the Pole so we took Crary and his assistant and took a bunch of measurements and so on. Then I stayed at T-3, but Bill went back with the airplane and when he got back, this, of course, gave him quite a bit of notoriety and it gave General Old a chance to intervene and get him off the hook.

(400)

BS: Tell me about the sweater incident. You're wearing a sweater.

JF: What:

BS: The sweater you're wearing here.

JF: Oh, well, that was. . . that sweater was knitted by a friend of mine who was one of my radar cohorts at the Radiation Lab and at Signal Corps labs. He'd worked for me during the war. His name was George Austin and he had been one of the four that I had taken with me to UCLA. While at UCLA, we naturally got acquainted with girls and George married one of them. I had introduced them and later on, when I got married, George was my best man, too. But, anyway, it turned out that Jeanie, the girl that he married, had knitted the sweater for me. And so this was a gesture to say that, yes, I appreciated the gift and wanted to show it at the time. And by the way, George Z. there is the parents of Tracy Austin who plays tennis. She was Number One for a while. Not too long. Navratalova beat her out. But there was a period of a few months when Tracy was Number One.

BS: So, tell me about the landing. The first landing at Pole. I understand you were back in the back and going to throw a smoke or . . .

JF: Oh, yes. Well, neither Bill nor I really were very good at judging ice thickness. And my idea that I favored was to land on refrozen lead because you're, you know, you're not so apt to hit a hump and break a ski and so on. Bill had even less experience than I, so he deferred to my judgment on that. So, anyway, we went up and we looked around in the immediate area. We had one of the navigators from the 58th B-29 squadron as navigator and we picked out a refrozen lead and were going to land on that. And we made a couple of low passes over it and we also had picked out on old floe which was right next to it, but which was really very . . . looked like it was much more chancy because it looked rough and we didn't know how rough.(450)

So, the upshot was that we made a couple of low passes over both of them, and then we decided that we would make the landing, but I wanted to put out a smoke flare. And so, I went . . . I was on the co-pilot's side. Bill was on the pilot's side and I wanted to toss a smoke flare out and I wanted to toss it at just the right time, so I'd be able to see it. And so I went to the rear of the airplane to toss out the smoke pot, or the bomb or whatever it's called, and Bill, then, was left in his seat. And as I said, he was kind of a wild guy, and damned if he, instead of making the low pass which we had agreed we were doing, he decided to go ahead and set down. So, it was a surprise to me whenever . . . he just cut the power and set it in on the floe, not the lead. And so I was at the back door with the door open at the time and I was mad as hell, at the moment. But, things went all right, so . . . .

BS: You were the first one out

JF: I was the first one out because I was at the door. BS: Oh, you fell out.

JF: I didn't fall out, but I jumped out. As soon as we came to a halt, I thought, hell, I might as go ahead and jump out. But, that's how I happened to be there.

BS: Any problems with getting off of the floe?

JF: We were . . . well we were both worried about it because it wasn't very . . . that's why we had been looking at the refrozen lead. And we had made a couple of low passes and on the second pass, we saw what I took to be a big pile of bear doo on the lead and it turned out that that isn't what it was. We had failed to reel in the training antenna and it was our trailing antenna that had pulled off and was laying there in the middle of the lead.

(500)

But, it's a good thing we didn't land on it because, on the lead. We did core through the ice and it was only about a foot and a half thick. Only about 15-16" and that might have been enough, but it might not, too. So that was a good thing that Bill landed on the floe.

BS: How did you nail the Pole, navigationally? How did you do that?

JF: We were only 100 miles away and Bill had brought up with him one of our best navigators from the B-29 squadron and so he was using, he was, of course, in daylight, and so all he had was the sun. But being only about, a matter of fact, it was only about 90 miles from T-3.

BS: The DR was pretty good.

JF: The DR was pretty damn good and that was backed up by his sun measurements and we had about 4 hours on the surface while we were there.

BS: And he confirmed it.

JF: He was taking sun measurements all during that time, so that's all we had. BS: Kept that record?

JF: I think so.

BS: Didn't register it with National Geographic or American Geographical Society?

JF: I don't think so.

BS: What was the science that Bert Crary and the others did? What did they do out there during that first period of weeks?

JF: Oh, Bert was there for 2-1/2 years. BS: Oh, he stayed for 2-1/2 years, on T-3?

JF: On the Ellesmere Shelf, on T-3, most of the time. On the Ellesmere Shelf part of the time. But, he was there almost continuously for the next 2-1/2 years and until he was appointed Chief Scientist for Antarctica. Or he was Deputy Chief Scientist. Harry Wexler was appointed Chief

Scientist and he appointed Bert as Deputy Chief Scientist for the IGY in Antarctica and I was madder than hell because, not that he isn't good for it, he was perfect. But, I was madder than hell to have him taken out of the Arctic. And Hugh Audeshaw, you remember him?

(550)

BS: Um-hum.

JF: Couple of guys, they flew up to Thule to interview Bert and to talk him into taking this appointment which they did without my knowledge. By that time, I was down in Maxwell and other duties, of course. But, anyway, they talked him into going to Antarctica. So he spent 2-1/2 years around T-3 and Ellesmere and then he was 2 years in Antarctica without ever going out.

BS: But what was the focus on science that he did? He and others - the science program.

JF: Well, really, that was wrapped around Bert. BS: Weather?

JF: No. No. Bert is an old seismographic. . . He was a geophysicist and his specialty was seismicity. He had worked for several years in Saudi Arabia and in the Persian Gulf just early at that time. He had worked in South America, and he had worked for the geophysics lab in Cambridge, and that's how he happened to be there because I had come from Cambridge. And of course, as soon as we started planning T-3, I called on the Cambridge bunch to provide the scientific component and I had met Bert before, but I was not well acquainted with him. So anyway, and Bert had had some experience the year before. Bert had been up to Alaska doing some seismic work around Barrow - he and his assistant.

(600)

And his assistant, which was carrying dynamite for him, what the hell was his name? He was later the President's Science Adviser, or no, the President of the Academy of Science, what the hell is his name? You know him.

BS: Phil Smith. Was Phil Smith his assistant? JF: Phil. No. He was never head of the . . . (End of Tape 2 - Side A

250

Two Photos from a late discovery

GRUMMAN AEROSPACE CORPORATION
• BETHPAGE, NEW YORK 11714'

ADDRESS REPLY TO:

GRUMMAN AEROSPACE CORPORATION
JOHN F. KENNEDY SPACE CENTER
KENNEDY SPACE CENTER ·
FLORIDA 32899

17 April 1970

Mr. T. J. O'Malley
Vice President
North American Rockwell
Kennedy Space Center, Florida   32899

Subject: SERVICES RENDERED

Inspection	$	20.00
Towing charge @ $1.00/mile		300,000.00
Loan of altitude vehicle .$20/day plus .08¢ per mile		24,100.00
Battery charge	・	5.00
Air conditioning @ $5.00/day		25.00
Room and board @ $40.00 each per day		600.00
	TOTAL:	$324,750.00

Very truly yours,

GRUMMAN AEROSPACE CORPORATION

George M. Skurla
Vice President

P. S. 2% contractor discount if paid within 30 days.

For Fed Haise

An ARPA ACV Funding Potential Never Materialized

C0-007-022 STK NO, J-2-020 REV \2 '66
INTEROFFICE MEMO.
AEROJET-GENERALCORPORATION
TO: V • J. Eggington DATE 11 November 1969
FROM: C. W. Ottinger
SUBJECT: Arctic Institute of North America A CV Interest
COPIES TO:
Encl. (1}
W. C. Beckwith, S. Eldridge, D. B. George, F. F. Herman, W. C House, Resume of Col. Joseph O. Fletcher
(USAF Ret.)

Joe and his wife have just returned from a ten-week working tour of Russia and were guests in my home November 9th. During his trip to Washington, D. C., during the past week he was appointed Chairman of a committee !or the Arctic Institute of North America which is preparing proposals for a research program to measure ice stress and contributing subsurface and above surface influences over large areas of the Arctic ice cap and over long periods of time. The utilization of ACV's for logistics support is highly attractive to the group and an ACV "workshop" session is planned by the Institute for December 15-16 in Massachusetts. Joe will be getting more details on the workshop this week but indicated the workshop was being funded by ARPA and that he would arrange an invitation to be given to me to attend. The resent phase for which Joe is chairman is to take the basic experiment proposal to the point where funds would be solicited to actually get hardware contracts underway. Some five or six manned stations will probably be involved in the experiment with that many or more unmanned stations. Distances of around 50 miles between each station are involved. I will continue to keep in touch with Joe and inform you as other information becomes available.

C. W. Ottinger